The Viewpoint Theory

Creation through the Eyes of the Creator

Daniel Baird

Copyright © 2019 by Daniel Baird
All rights reserved.

ISBN: 9781687195968

Imprint: Independently published

Contact Daniel Baird at: theviewpointtheory@gmail.com

Contents

Introduction   page 1
  The Viewpoint Theory

Chapter 1   page 7
  A Leap of Faith

Chapter 2   page 19
  The Genesis Creation Story:
  A Personal Approach

Chapter 3   page 31
  The Mind of God

Chapter 4   page 35
  The Three Essential Insights

Chapter 5   page 45
  7th Day—The Natural World

Chapter 6   page 57
  The Unseen Source

Chapter 7   page 67
  The Three Fundamental Principles

Chapter 8   page 75
  Writing the Genesis Creation

Chapter 9 page 89
The Divine Order

Chapter 10 page 121
The Search for Meaning

Chapter 11 page 135
If God Created the Universe

Acknowledgements page 149

Introduction

The Viewpoint Theory

"What I am looking for is not out there, it is in me."

–Helen Keller

As I breathe in the cool ocean air, I marvel at my own existence. I am here, experiencing this magnificent world. How can that be? I am not alone in my wonder! The mysteries of being and creation have sparked the human imagination for thousands of years.

The telescope tells us that the scope and history of the universe is immense. Its story is so big that it can make us feel small and insignificant. Ultimately, however, all reality may not be as it seems. The secret that makes a magic trick appear real is not what you see, but is what you don't see. The telescope does not see the spirit!

> *"Where the spirit does not work with the hand, there is no art."*
> –Leonardo da Vinci

Many of us begin to question reality and the meaning of life in early childhood. I remember being 4 years old and wondering, "Why am I here?" As we grow older, distraction can become a habit, and so the mysteries of life are pushed to the back shelf. But for a few of us, an insistent voice inspires us to go deeper into the unknown.

Searching the depths of meaning and creation alone can be a daunting task. It can feel like being lost in a house of mirrors. Every way you turn is your own reflection, until eventually; you find the exit door of religion or science. To end the pain of confusion, many of us will inevitably reach for one or the other.

Most religions, like the ancient Egyptian and Greek religions, Hinduism, paganism, Judaism, Islam, and Christianity, own a creation myth. They lay claim to the true story of the universe and us. The religious narrative has dominated human consciousness throughout the ages. Science, though, is slowly starting to gain control over human thought. It is becoming the authority!

Unfortunately, for religion's sake, science has been telling a different story. The fields of biology, physics, geology, and astronomy reveal a complex picture of creation. The sciences unveil an evolving universe that makes religions stories seem like pure fantasy. In the academic world, ancient creation accounts, like Genesis, are not seen as offering a serious explanation for the universe.

> *"You've never read Genesis? You should... It's hilarious."*
> *—Atheist humor*

Religion's influence over our mind is fading, primarily due to science. Still, there are religious groups which are not going down without a fight. Creationism uses science and reasoning to furiously defend Genesis against the attacks of atheists and others. Atheists assert that a supernatural creator does not exist. They believe science shows that all reality is solely the manifestation of natural phenomenon. Creationists insist that God is real. They believe that God is the creator of everything, and science clearly supports this fact.

Does science prove or disprove a creator? The conflict over this very question has escalated in recent years. But, the war has been a little one–sided. The majority of scientists, especially the most famous ones, adamantly oppose the ideas presented by creationists.

"Before we understood science, it was natural to believe that God created the universe, but now science offers a more convincing explanation."

—*Steven Hawking*

Science is continually unraveling the mysteries of the universe. Scientists offer natural explanations to things once believed to be only explainable supernaturally. A creator has become, in the minds of many, completely unnecessary.

"The lack of understanding something is not evidence of God; it's the evidence of lack of understanding."

—*Lawrence Krauss*

The prediction that science can explain all truth without needing God poses a threat to faith. One is forced to face the question. "How can you believe in science and God simultaneously?" Many young people are rejecting God based on the inability to reconcile the two. *The Viewpoint Theory* reconciles a creator and science without compromising either one.

You need science to explain the universe, but you need God for there to be a universe that science can explain.

I have always had a longing for a deeper meaning to the universe and life than what traditional science is offering. *The Viewpoint Theory* gives the spirit a voice in existence, reality, and creation. The Genesis creation is a story based in the language of spirit. Hidden at the heart of it beats a divine order, which can only be heard through your soul!

1

The Viewpoint Theory

A Leap of Faith

Jeremiah 33:3 "Call to me and I will answer you and tell you great and unsearchable things you do not know."

To have doubt or to have faith–which strategy unveils the real creation of the universe? Doubt is adopted by science, and faith is embraced by religion. Doubt is crucial for scientific inquiry. Without it, science would not work or evolve. If you didn't question reality, how could you discover anything? Questioning the nature of the world, and being skeptical of the answer makes for a great scientist.

The tactic of doubt has led to countless scientific and technological discoveries. We are completely reliant on the comfort and security they bring us. Science and technology have become the problem solvers of the new millennia. Our total dependence on them however, has twisted them into something they are not. Science is not God!

> *"Science is a way of thinking much more than it is a body of knowledge."*
> –Carl Sagan

Does the scientific method and doubt guide you to an objective and unlimited view of all reality? I doubt it! Doubt has limits on how far it can take you! If you push doubt to its limit, it turns in on itself. The moment you become the skeptic of your own skepticism, or the doubter of your doubt, faith is awakened.

Religion often binds faith with dogma. Faith that is free from rules is dynamic. It never ceases searching the nature of God. Faith is essential in transcending limitations. Having faith in God, without a cause or a reason, allows your thoughts to escape the gravity of past knowledge. Only then can you see into the beyond.

Mark 9:23 "... Everything is possible for one who believes."

Science and doubt cannot validate or invalidate God. They haven't so far! If they could, it would defy God's nature. A creator's very existence means that science is only offering a partial view of reality.

There is a box of rules for reality which have been established. If you think inside the box; you are considered reasonable and a realist. The unreasonableness of faith is the only way to break free and explore outside the box.

Atoms, particles, elements, chemicals, matter, antimatter, and so on are the stuff of science. Everything that is more than science is the stuff of faith. If God created the universe, the answer won't be found through the fields of science. You must go straight to the source. This calls for a leap of faith.

Finding the Creator

1Corinthians 2: "The spirit searches all things, even the deep things of God..."

The Viewpoint Theory is an exploration into the infinite potential of reality. It envisions viewpoints that transcend every possible boundary and law. The word in English, which exceeds all limits, is *God*.

All other words fall within the structure of our universe and take on a restrictive definition. *God* is limitless Being! Embracing *God* creates the ability to venture into the unknown and explore the foundation of everything.

> *"God is a metaphor for that which transcends all levels of intellectual thought..."*
> –Joseph Campbell

The raging debate between atheists and creationists is based in an agreement. The agreement is that scientific evidence can be found, which proves or disproves a creator. Both sides agree that the universe is a reference for reality, or truth, from which you can question the existence of God.

> *"Deep down, creationists realize they will never win factual arguments with science. This is why they have construed their own science–like universe..."*
> –Frans de Waal

> *"There is not the slightest possibility that the facts of science can contradict the bible..."*
> –Henry Morris

The Viewpoint Theory begins from the existence of God, and then questions reality through that medium. When you imagine looking through the eyes of a limitless creator your perception expands.

This act, however, does create a paradox. You notice that our reality, which seems to be absolutely true, is an illusion from a greater viewpoint. Yet, as the sciences discover more and more each day, it is a supreme God that is judged as the illusion.

> *"The more I learn about the universe, the less convinced I am that there's any sort of benevolent force that has anything to do with it at all."*
> —Neil deGrasse Tyson

We invented "God" in our minds is a popular intellectual belief. God and gods have been fantasized since ancient times as a way to cope with and explain unknown natural occurrences. Fire, floods, earthquakes, and celestial events must have seemed supernatural to primitive people. Modern religions are simply perpetuating an archaic fairytale. So the only rational conclusion is that god is a delusion.

The intellect can dismiss God easily, but the spirit cannot. It is impossible to create God with the human brain. Our thinking is very clever at inventing opinions and beliefs about God, but is powerless in touching the eminence of God's nature.

> *"...The intellect cannot get the illumination of God's wisdom, the will cannot get the love of*

God, and the memory cannot get God's image..."
               —*St. John of the Cross*

If you want to understand someone, walk a mile in their shoes. The wisdom of this well–known proverb, if practiced, would likely solve most of the problems in the world. The empathy it inspires allows you to feel other perspectives. This is the secret to understanding *The Viewpoint Theory* and the Genesis creation story. Embracing the viewpoint of an unlimited being, creator and observer opens you to limitless realities. Imagine the viewpoint of an unlimited creator! From that point of view, seek the wisdom of creation.

When you factor God into creation you get much more than the universe.

The famous physicist, Albert Einstein, allegedly said, "No problem can be solved from the same level of consciousness that created it." A very intriguing question can be formed from this quote. Is the level that we perceive reality capable of solving the mystery of creation? The natural sciences only resolve problems that are on the same level as the known universe. We must visualize reality through a higher viewpoint if we truly want to glimpse the whole picture of creation.

Trying to explain the true origin of the universe with the universe or anything that is a product of it is like a dog chasing its own tail.

Great philosophers, thinkers, sages, and religious figures through the ages have all proposed the same fascinating insight. They believed that there is an interconnection between the perceiver or observer, and the reality which is observed or witnessed. Some stretch this idea further, and suggest that one's consciousness and perception influences or even creates reality. Contemporary quantum physicists seem to encounter this same or similar notion—the act of pure observation affects reality.

> *"Science cannot solve the ultimate mystery of nature. And that is because, in the last analysis, we ourselves are part of the mystery that we are trying to solve."*
> –Max Planck

The Viewpoint Theory is founded upon the principle that what appears to be true or real from a human viewpoint would not be the same as what is true or real from the viewpoint of an unlimited observer. It considers what reality and understanding would be like through the mind of a limitless creator. What would it truly mean if God created the universe?

A scientist relentlessly pursues natural theories for truth and reality. The Viewpoint Theory relentlessly pursues the inner nature of God.

The science fiction movie, Contact, which was released in the 1990s, presented a clever anecdote to illustrate the connection between understanding and viewpoint. In the movie, the leading character, played by Jodie Foster, receives a radio signal containing a math code for how to build a vehicle which could transport a human across the galaxy to an alien civilization. The only problem, however, was that none of the mathematicians could decipher the code. Eventually, someone had the brilliant realization that the reason they could not fully translate the alien transmission was because they were analyzing it from a human viewpoint. They recognized that they needed to think as an advanced alien race would think if they were to crack the code. So when they imagined themselves in the mind and viewpoint of a highly advanced alien civilization, it expanded their consciousness, and allowed them to decode the message.

If God created the universe, for us to have some kind of understanding of how God did it, we would need to do our best to view everything as an eternal, unlimited, omnipotent, omniscience, and omnipresent being.

Warning! Do not be tempted to try and entrap the creator with physical laws and facts. The only thing that will get caught is your own thinking. Your thoughts will be forced to obey the laws and facts you hold true. You will find that there is no need to make the creator compliant with observable scientific evidence. God is real without it!

> *Luke 17:20 "The kingdom of God does not come with what is observed."*

Evidence that is found from visible, changing, limited, and caused observations is powerless in revealing the existence of an unlimited, unchanging, invisible, and uncaused God.

I expect that many people would argue that it is impossible to imagine God's viewpoint. Not to mention, glimpsing into how God created the universe. They might say God could create the universe however God wants to. There is an important distinction to be made here between how a creator *could* create the universe and how a creator *would* create the universe. Certainly one could reason that God is unlimited. So therefore, God could create the universe however he chooses to. *The Viewpoint Theory* proposes that God *would* create the universe in integrity with his own nature.

The Genesis creation story is based in God creating in faith with his nature. The author repeatedly states,

"God saw his creation that it was good." This clearly demonstrates that God's creation is compatible with his supernatural viewpoint and is integrated with his word.

> *John 1:1 "In the beginning was the Word, and the Word was with God, and the Word was God..."*

The Viewpoint Theory was not originally discovered by researching the Genesis creation story. Nor was it ever intended to be an interpretation of it. It all began when I was looking out over the ocean one day and wondered, "If God is the creator; how did we end up with this world? If there is being and reality which transcends time, how did we get from there to here or from that to this?" These questions led to the discovery of The Divine Order—a timeless foundation from which our universe arises. It just so happened that The Divine Order seemed to perfectly coincide with the 7 days of the Genesis creation.

Most interpretations of the Genesis creation story try to understand it within the framework of the past. It is studied as a manuscript pertaining only to historical cultures, characters, and events. *The Viewpoint Theory* perceives this ancient story completely through the power of the present. The Divine Order can be discovered free from the Genesis creation story. So the true intent behind its narrative is to reveal a sacred order, which is eternal and present now. Therefore, it is available for anyone to find!

2

The Viewpoint Theory

The Genesis Creation Story: A Personal Approach

Mathew 6:33 "But seek ye first the kingdom of God and his righteousness..."

You're crazy! That's a look I get when I tell others about my theory of everything. They may not say it aloud, but I imagine them secretly thinking, "Who do you think you are? Where did you get your college degree--Looney Tunes University? You don't have the authority to make such a claim." My personal story may help clarify what guided me to this theory. It shows why having a free, original, and intuitive mind is essential to understanding the Genesis creation.

As a small child in school, I would often experience what I now know as anxiety attacks. My palms would leave puddles of sweat on the desk. My hands would shake, heart race, and face would get flushed. At that time, I had no idea what was happening to me. So I never told anyone about it. These panic episodes went on from kindergarten all the way through high school.

In my first year in college, the panic attacks seemed to rise to a new level. They began to make me feel dizzy and disoriented. Depression started to sink in. I felt like I was living in a paralyzed state! The anxiety became so great I did not feel like I could even function in society. So I dropped out of school and avoided the world.

For several years, into my early twenties, I worked at the type of jobs that afforded me income and the ability to escape people as much as possible. I had no ambition. Life seemed rather meaningless. Success equaled having to endure panic attacks. So why would I want that? Eventually, I reached a breaking point. I could turn to drugs or seek a spiritual meaning to life.

I remembered my brother reading to me, as a teenager, an Indian mystic named Jiddu Krishnamurti. His teachings resonated with me. The idea of being free from the traps of thinking was very appealing to me. If I could quiet the thinking mind, perhaps it would relieve my anxiety. This was my hope. So I started on my spiritual journey, walking the beaches in Daytona, Florida.

Every day I would wake up with the quest to seek the truth. I would spend hours walking the beach observing nature and watching every thought and reaction of my mind. About a year or so into this practice, I started to have extraordinary experiences. I would experience my consciousness, or being, expanding out beyond my brain and body—an amazing feeling, beyond words, of bliss, peace, love and connection with everything.

While in these expanded states, I seemed to understand life and the world very differently. I began to intuitively feel that there existed even greater realities–viewpoints, which transcend the known universe. I sensed that our universe really arises from beyond the reach of our intellect.

For the next several years, I organized my insights into what I call The Divine Order. It wasn't until later, when I received a King James Bible as a Christmas present that I realized The Divine Order seemed to also be the foundation of the Genesis creation story.

The Genesis Creation Story

Those of us educated within western cultures are familiar with the Genesis creation story. It is a creation narrative, believed to be written by Moses, adopted by Judaism and Christianity. You have probably read it in the Christian Bible or the Hebrew Bible at least once, voluntarily or involuntarily. If you were captivated enough by it, you may have decided to compare these ancient writings with contemporary knowledge. Perhaps you were moved to contrast it with scientific observations of the world. You no doubt soon discovered, however, that big problems arose when comparing and contrasting the formal biblical words with modern science.

It becomes quite noticeable when reading Genesis that it does not possess the current understanding of physics, chemistry, astronomy, and biology. The author did not own tools like telescopes or microscopes necessary to fully comprehend the magnitude, complexity, and ongoing processes of the universe and nature.

> *"Our world is very different from the world in which the Biblical authors lived over 2,000 years ago. The ancient world did not have Google, Wikipedia, and smartphones- access to information on human history and scientific achievements developed over millennia at the touch of their fingertips."*
> *–Shawna Dolansky*

To cast further doubt onto the veracity of the Genesis creation story, some historians claim it is based primarily in ancient myths dating back many years before its inception. They also believe it was written by more than one author, living centuries apart, and is comprised of two creation accounts. Some scholars theorize that the motivation for writing the Genesis creation was restricted to the author's culture. They conclude it is not describing world historical events, but is primarily a story taken from Babylonian mythologies.

There are many scholars and historians however, that do claim that Genesis was written by Moses, as the Bible states. Still, regardless of who wrote the Genesis creation or where it originated from, its essence is not based in an ancient author's subjective and exclusive tale. Rather, it is founded in an objective and inclusive self–evident revelation.

"Faith is reason plus revelation, and the revelation part requires one to think with the spirit as well as with the mind..."
—Francis Collins

Science is a word that is often associated with objective and non-personal truth, while revelation is linked with only subjective and personal beliefs. But true revelation comes through the spirit, which connects all of us.

We are all exactly the same, in that we are individual souls. Revelation is spiritual truth, which means it is undivided from others.

Did the people in ancient civilizations all hold the same beliefs about the universe? Were their thoughts trapped in the culture, knowledge and myths of the time period? I suggest there have always been free thinkers!

History is filled with visionaries who could see the world in a brighter light. The Genesis creation was an inspired narrative. Its essential meaning is not limited to an ancient culture. It uses the literary devices of allegory and metaphor to reveal God's unseen, eternal and divine creation.

Mathew 13:35 "I will open my mouth in parables; I will utter things hidden since the foundation of the world."

The 6 days of the Genesis creation may appear to be all about the physical world that we see, but it is really about the spiritual world that we don't see. Therefore, the treasures of its words can only be received on a soul level. Forcing it to make logical sense on a material level is futile. When you search for its purpose within your heart, its virtue will become apparent.

> *"The universes which are amenable to the intellect can never satisfy the instincts of the heart."*
> –Anonymous, The Cloud of Unknowing

Perhaps our predisposition to think that the Genesis creation can be comprehended through the academic disciplines of history, biology, psychology, geology, and astronomical evidence is incorrect. *The Viewpoint Theory* suggests that the Genesis creation was not an attempt at natural science. The author wasn't trying to teach biology or physics. The story describes the relationship dynamic between creator, spirit, light, consciousness, mind, change, creation, universe, life and us.

**The Biblical Creation Story:
Genesis 1 and 2**
(King James Version of the Bible)

1 In the Beginning God created the heaven and the earth.

2 And the earth was without form, and void; and darkness was upon the face of the deep. And the spirit of God moved upon the face of the waters.
3 And God said, let there be light: and there was light.
4 And God saw the light, that it was good: And God divided the light from the darkness.
5 And God called the light day, and the darkness he called night. And the evening and the morning were the **first day**.
6 And God said, let there be a firmament in the midst of the waters, and let it divide the waters from the waters.
7 And God made the firmament, and divided the waters which were under the firmament from the waters which were above the firmament: and it was so.
8 And God called the firmament heaven. And the evening and the morning were the **second day**.
9 And God said let the waters under the heaven be gathered together unto one place, and let the dry land appear: and it was so.
10 And God called the dry land earth; and the gathering together of the waters he called seas: and God saw that it was good.
11 And God said, let the earth bring forth grass, the herb yielding seed, and the fruit tree yielding fruit after his kind, whose seed is in itself, upon the earth: and it was so.
12 And the earth brought forth grass, and herb yielding seed after his kind, and the tree yielding

fruit, whose seed was in itself, after his kind: and God saw that it was good.

13 And the evening and the morning were the **third day.**

14 And God said, let there be lights in the firmament of heaven to divide the day from the night; and let them be for signs, and for seasons, and for days, and years:

15 And let them be for lights in the firmament of heaven to give light upon the earth: and it was so.

16 And God made two great lights; the greater light to rule the day, and the lesser light to rule the night: he made the stars also.

17 And God set them in the firmament of heaven to give light upon the earth,

18 And to rule over the day and over the night, and to divide the light from the darkness: and God saw that it was good.

19 And the evening and the morning were the **fourth day**.

20 And God said, let the waters bring forth abundantly the moving creature that hath life, and fowl that may fly above the earth in the open firmament of heaven.

21 And God created great whales, and every living creature that moveth, which the waters brought forth abundantly, after their kind, and every wing fowl after his kind: and God saw it was good.

22 And God blessed them, saying, be fruitful, and multiply, and fill the waters in the seas, and let fowl multiply in the earth.

23 And the evening and the morning were the **fifth day**.

24 And God said, let the earth bring forth the living creature after his kind, cattle, and creeping thing, and beast of the earth after his kind: and it was so.

25 And God made the beast of the earth after his kind, and cattle after their kind, and everything that creepeth upon the earth after his kind: and God saw that it was good.

26 And God said, let us make man in our image, after our likeness: and let them have dominion over the fish of the sea, and the fowl of the air, and over the cattle, and over all the earth, and over every creepy thing that creepeth upon the earth.

27 So God created man in his own image, in the image of God created he him; male and female created he them.

28 And God blessed them, and said unto them, be fruitful, and multiply, and replenish the earth, and subdue it: and have dominion over the fish of the sea, and over the fowl of the air, and every living thing that moveth upon the earth.

29 And God said, behold, I have given you every herb bearing seed, which is upon the face of all the earth, and every tree, in the which is the fruit of a tree yielding seed; to you it shall be for meat: and it was so.

30 And to every beast of the earth, and to every fowl of the air, and to everything that creepeth upon the earth, wherein there is life, I have given every green herb for meat: and it was so.

31 And God saw everything that he had made, and, behold, it was very good. And the evening and the morning were the **sixth day.**

1 Thus the heavens and the earth were finished and all the host of them.
2 And on the **seventh day** God ended his work which he had made; and he rested on the seventh day from all his work which he had made.
3 And God blessed the seventh day, and sanctified it: because that in it he had rested from all his work which God created and made.
4 These are the **generations** of the heavens and the earth when they were created, in the day that the lord God made the earth and the heavens.

3

The Viewpoint Theory

The Mind of God

Proverbs 3:5 "Trust the Lord with all your heart and lean not on your own understanding."

If you wanted to attract a lot of attention in ancient times, just proclaim that you are talking with God. The belief that the author of Genesis was communicating directly with God has enticed renowned historians, saints, priests, and scholars for centuries. All these great minds trying to figure out this ancient epic has led to many differing theories and opinions.

The Definition of viewpoint as "a mental position from which something is viewed" serves the purpose of *The Viewpoint Theory*. So in essence, it is through the "mind of God" that the six days of the Genesis creation story is expressed and perceived. Notice that the last verse in Genesis 1 proclaims it is through the viewpoint of God that all creation is being observed.

> *Genesis 1:31 "And God saw everything that he had made, and, behold it was very Good."*

Attempting to explain the Genesis creation through a human viewpoint is a fatal mistake. The human brain is interrelated with the natural world around us. We perceive reality, primarily through time and space, causes and effects, and physical change. Our brains don't seem to be hardwired to perceive reality beyond these parameters. So we project our foundations for reality onto translating the ancient verses in Genesis. This leads to a limited and incomplete understanding.

The source of everything is not in human perception. It is within the omniscience of God.

There is a common thread that connects the many Genesis creation theories. They may seem on the surface to be diverse. But, if you look a little deeper, you will see that most are based on our human reality. They comprehend the story through time and space, causes and effects, and physical change. They frame it into a narrative which makes logical sense to our minds but not to the mind of an omnipresent being.

The Genesis of creation is not in the absence of the past. It is within the omnipresence of the creator.

Underneath the narrative of the Genesis creation there is a divine order (Chapter 9). The Divine Order is intuitively uncovered only when you appreciate Genesis through the mind that authored it-God. I don't discount historical, religious and scientific perspectives on the Genesis creation story. They all have value! But, they do not reveal The Divine Order.

We expect to find a logical cause to everything, but the fullness of God's creation cannot be known through our thinking mind.

The "mind of God" perceives through the incomprehensible principles of absolute Goodness, perfection, love, truth and limitlessness. If the wisdom of the Genesis creation is to be truly realized, we need to accept that it was written from the viewpoint of these higher principles.

> *Isaiah 55:8-9 "For my thoughts are not your thoughts, nor are your ways my ways," declares the Lord. "For just as the heavens are higher*

than the earth, so are my ways higher than your ways, and my thoughts than your thoughts."

God's creation is not in the limits of the observed. It is within the limitlessness of the creator.

Approaching the Genesis creation with unwavering faith in God's mind and nature opens the door to a new way of understanding it. Seeking its meaning, while being intently focused on the creator, unveils the sacred creation!

4

The Viewpoint Theory

The Three Essential Insights

Luke 17:21 "Nor will people say, 'here it is', or 'there it is' because the kingdom of God is in your midst."

Are you ready to go down the rabbit hole? Profound dimensions lay deep within the Genesis creation story, which have not yet been explored. There is a buildup of dark confusion that surrounds it. Insight can bring light into the darkness. The following are three essential insights to help guide you into its deeper realms.

The First Essential Insight–Free Mind

To begin, take a deep breath, relax, and expand your mind. Recognize that you don't need to be anybody else. You don't need science or the possession of secret or special knowledge. You only need to ask God and be open to receive.

> *Luke 11:9 "Ask and it will be given to you; seek and you will find; knock and the door will be opened to you."*

Our first impulse when we engage a challenging subject like the Genesis creation is to research the work of others. We may feel like we must become someone else, if we are to comprehend it. We might feel unworthy of profound insight, if we don't have a Ph.D. in a field like archeology, history, biblical studies, or mythology. Since most of us have not studied that extensively, we resign ourselves to just accepting those that have acquired the education as the authority.

Perhaps in almost every instance, it is wise to seek the council of experts before venturing into a mystery. This is not one of those instances! In the case of Genesis, it is vital to begin with a silent and clear mind—a mind that is unclouded by any point of view. Seeking God's creation through the opinions, facts and evidence of authorities keeps you from being receptive to something greater.

> *"It is best to learn to silence the faculties and to cause them to be still so that God may speak."*
>
> *–John of the Cross*

The telescope, microscope, carbon dating, and Hadron Collider have been proven to be highly effective in answering the questions of physics, astronomy, biology, and history. Yet these are not instruments and methods sufficient to unravel the subtlety of the Genesis creation. It is your inner gifts of awareness, consciousness, imagination, insight, intuition, revelation, faith, and perception that bring lucidness.

> *"I believe in intuition and imagination…I sometimes FEEL that I am right. I do not KNOW that I am."*
>
> *–Albert Einstein*

Using telescopes to look far out into the cosmos and examine the relationships and behaviors of planets, distant stars, galaxies, black holes and beyond is magnificent for discovering the nature and origins of our physical world. Even the most powerful telescope, though, cannot pierce into the origins and nature of the Genesis creation. Observing within and receiving spirit–knowledge is its true home.

The 2nd Essential Insight–The Ultimate Vision

"The ability to perceive or think differently is more important than the knowledge gained."

–David Bohm

The Genesis creation story presents the ultimate vision of reality when observed through the viewpoint of a limitless creator. From a modern point of view, the Genesis creation appears to be an analysis of the world which is archaic. It seems to be an antiquated work, proven to be illogical and flawed. The notion that astronomy, physics, and biology expose a much bigger, more complex, and superior universe than the Genesis creation is a complete misunderstanding of it. This misinterpretation has caused many people to disbelieve in it and God.

To truly receive a deeper glimpse into the immensity, perfection and aliveness of the Genesis creation, one needs to embrace it wholly through the newness of the present. If you mentally lock it up in the distant past, you will perceive it as imperfect, old, and dead.

Is the Genesis creation story meant to be interpreted literally? A very contentious debate has been raging for many years over this question. *The Viewpoint Theory* suggests that this masterpiece is absolutely literal, but only from the creator's viewpoint.
Our human language and observations of nature are relative, partial, and restricted. Therefore, they can only serve as descriptions of God's literal creation. If you relate to the Genesis creation as a creator, you will recognize its words are meant to be treated metaphorically. Our words, by their very nature, are symbols and images within our minds that represent things, meaning, and reality. The metaphors, symbols, and images utilized in Genesis to reveal God's nature, The Divine Order, and God's intentions certainly cannot literally capture the incomprehensible truth of those realities.

> *Psalms 147:5 "How great is our lord! His power is absolute! His understanding is beyond comprehension!"*

The 6 days of Genesis uses our normal everyday images, impressions, observations, relationships, and experiences like male and female, light and darkness, day and night, rest, sight, sun, sky, earth, animals, plants, and water metaphorically to reveal the unseen creation. The writer carefully expresses and organizes our common human impressions to unveil God's true creation. The fascinating thing is that the trees, birds, stars and everything else we perceive in this world are literally themselves relative imitations, images, and descriptions of God's authentic creation. This will be discussed in more detail in the next insight.

The Viewpoint Theory offers the vision that the entire universe and its content, which we see and know, are but incomplete reflections and images of the full universe God sees and knows.

> *Corinthians 13:12 "For now we see only an indistinct image in the mirror but then we will we be face to face. Now what I know is incomplete, but then I will know fully, even as I have been fully known."*

The 3rd Essential Insight–God's Mind Versus Human Mind

The first 6 days in Genesis reflects the *mind of God*, while the 7th day mirrors our *human mind*. This insight unveils an extraordinary realization. The universe that we are familiar with is not the same as God's creation.

People of faith often gaze into the beauty of nature and say something such as, "This is God's wonderful creation," but that may not be entirely true. It may be closer to truth to look at the magnificent things in nature, and say, "These are wonderful impressions or images of God's creation."

Our universe is like a motion picture of God's creation, which we help project. Everything that we believe is solid is really just the movement of perceptions, images and impressions.

> *"What quantum physics has taught us is that everything we thought was physical is not physical."*
> *−Bruce H. Lipton, PhD*

When you look at a tree, you may think that you are observing or receiving an impression of something that is physical or real. But the tree itself is also an impression that is being received by the world. This insight is a vital key to understanding *The Viewpoint Theory* and the Genesis creation.

The creator's true or real creation is unveiled in the 6 days of Genesis. It is not witnessed through the observations of science.

> *Luke 17:20 "...The Kingdom of God comes not with observation."*

The images, impressions and memories of God's creation are revealed on the 7th day. It is the universe that we play in and that science comprehends.

The sheer vastness, unfathomable power and beautiful firework displays of the cosmos, hypnotize us into believing that we are witnessing God's creation, when in truth, it is merely the reflections and memories of it.

God's creation is much greater than the universe. Think of our universe as a subjective experience that we are all having together. We may believe that gravity is absolutely true, but it does not govern or rule God's creation. The laws of science cannot relate to it, which means they are ultimately not absolute or objective.

> *"I think science has enjoyed an extraordinary success mainly because it has such a limited and narrow realm in which to focus its efforts. Namely, the physical universe."*
> –Ken Jenkins

Our human brain is not capable of beholding God's actual creation. We can acknowledge, however, the images, impressions, and memories of God's masterwork. They reflect the nature of our minds.

Carefully read the 6 days of Genesis, and you will notice that it is conforming to God's viewpoint. On the 3rd, 4th, 5th, and 6th days it states, "God saw (not humans saw) his creation that it was Good." The expression "God saw" demonstrates that the creation is responding to God's viewpoint. The word "Good" shows that the creation is reflecting his nature. At the finish of the 6th day, it reads,

> *Genesis 1:31 "And God saw everything that he had made, and, behold, it was very Good."*

Subsequently, in *Genesis 2:1*, it proclaims *"thus the heavens and the earth were finished, and all the host of them."*

Notice! The author very clearly articulates in Genesis 2:1 that God had finished creating the heaven and the earth. God's work was done! There is an important shift in perception that begins with Genesis 2:2.

> *Genesis 2:2 "And on the 7th day God ended his work which he had created and made; and he rested on the 7th day from all his work which he had created and made."*

God resting on the 7th day has a special significance. The word "rested" in Gen 2:2 is a metaphor to convey a feeling of receptiveness and inaction. The shift from the 6th to 7th day is signifying that *the mind of God* is transitioning from a creative state to a receptive state. So God "resting" symbolizes the creator releasing his viewpoint hold on creation. This releasing allows reality to acknowledge our *human mind*. Notice the next verse:

> *Genesis 2:3 "And God blessed the 7th day, and sanctified it: because that in it he had rested from all his work which God created and made."*

The creator is letting go of his creation–like a father letting go of his child's bicycle with blessing and faith that the child will petal and balance the bike on her own.

The next verse, Genesis 2:4, is a key to validating *The Viewpoint Theory*.

> *Genesis 2:4 "These are the generations of the heavens and of the earth, when they were created, in the day the lord God made the earth and the heavens."*

The heavens and earth are now moving and changing on their own like the child riding her bike without her father controlling it!

5

The Viewpoint Theory

7th Day—The Natural World

Ecclesiastes 3:1 "There is a time for everything, and a season for every activity under the heavens."

Have you ever wondered why the Genesis creation ends at 7 days? Why is there not an 8th day or a 9th day? It is because the 7th day is the realm of the natural world. It is where linear time or forms changing becomes a reality.

God creates or causes time and change, but they do not reflect his nature.

> *Malachi 3:6 "For I am the Lord, I change not...."*

Amazingly, the natural world and things changing arises through the inaction of the creator. This is why God seems so irrelevant to science. How do you find the evidence for inaction? Yet, God resting or entering into a receptive state on the 7th day is absolutely necessary for the natural world to exist and for us to experience generations, time, days, years, and centuries.

The artist must step away from her masterpiece if she wants others to experience or interact with it.

The Viewpoint Theory proposes that God endowed his creation with the potential for it to be creatively expressed through the natural world. The universe, with all its galaxies, grains of sand, and life, is in a continuous process of creatively replicating God's creation.

> *"The universe is not in a steady state; there's an ongoing creative principle in nature, which is driving things onwards."*
> *–Rupert Sheldrake*

The principles of science become relevant on the 7th day. So if the Big Bang theory really happened, it is only true from the viewpoint of the natural world or the 7th day.

Genesis 2:4 *"These are the Generations of the heavens and the earth,"* tells us that we have entered the realm of linear time (past, present and future). And that we are witnessing the generations, images, or impressions of the heavens and earth.

Genesis 2:4 has challenged bible scholars for centuries. This verse does not seem to neatly fit, at least from a human viewpoint, with the first 6 days of God's creation. It doesn't fit because God's creation is *finished*! The natural world is the generations of the creation. It is an *unfinished creative* reproduction.

Genesis 2:5–2:7, in the 7th day, describes entering into the domain of the natural world. *"And every plant in the field before it was in the earth, and every herb of the field before it grew, for the lord God had not caused it to rain upon the earth and there was not a man to till the ground. But there went up a mist from the earth, and watered the whole face of the ground. And the lord God formed man of the dust of the ground, and breathed into his nostrils the breath of life; and man became a living soul."*

In God's 6 days of creation, reality is not caused, as someone throwing a rock in a pond causing the water to ripple. Everything is created instantaneously, directly, and purposefully through and by God. In contrast, Genesis 2:5–2:7, describes reality that has transitioned into the natural world, wherefrom the Lord causes substance to interact with itself in order to generate change. This is the realm of physics and biology–where plants and trees interact with the soil, water, elements and sunlight to cause them to grow, reproduce and multiply. The natural world can express Gods creation in infinite creative ways.

> *"The creation of a thousand forests is in one acorn."*
> *–Ralph Waldo Emerson*

God's deliberate creation of everything we see like the stars, the sun, and earth, become natural occurring events on the 7th day. Man, who was created in the image of God on the 6th day, becomes manifested through biology brought to life through the ground, dust, elements, and air (or oxygen) on the 7th day.

The author begins to refer to God as Lord on the 7th day. This indicates that the 7th day is written from a human viewpoint. The word, Lord, illustrates that the writer's viewpoint is subject to God and his creation. The author is now perceiving and experiencing God and the creation through the medium of the natural world.

Some biblical historians claim that Genesis 2:4 *"These are the generations of the heavens and the earth,"* is the beginning verse to a completely separate creation story, which was written centuries before Genesis 1.

There is a sacred connection between the 6 days and the 7th day of creation which cannot be broken. So even if Genesis 2:4 was inscribed by an author from a different time, it would not alter this truth. There is a divine order that interconnects the 7 days. It eclipses historical inconsistencies and overcomes any discrepancies in translations. It transcends the limitations of human language and bias.

The English word, *generations,* is used by The King James Bible to translate an ancient Hebrew word in Genesis 2:4. Different words have been used as translations by other bible versions, such as *histories*, *accounts*, *records*, and *origins*. These word deviations are an indication that Genesis 2:4 has not been fully understood. Therefore, attempts are made to force it to logically fit into our human perspective.

The author of Genesis 2:4 is conveying that the heavens and the earth are going through a process of change. The word, *generations*, seems to encompass that idea in the best way.

Presented earlier was the concept that our reality is based primarily on space and time (past, present, future), causes and effects, and physical change. Any event that we seek to comprehend, our first instinct is to orientate it with our foundation for reality. When and where did it happen, what caused it to happen, and what change occurred as a consequence, are essential questions that we innately ask.

If we desired to learn about the fate of the dinosaurs, we may begin by asking, "When and where did they become extinct?" The consensus is, "On earth about 65 million years ago." The next question may be, "What caused the dinosaurs to become extinct?" It is believed that an asteroid colliding with the earth sealed their fate. Lastly, to gain further knowledge, we may ask, "What changed as a consequence from this event?" The extinction of the dinosaurs allowed for new life to emerge.

If we pay close attention to our daily life activities, for example, getting a haircut, pumping gas into our car, playing a sport, learning in school, paying the bills, and going to work, it becomes apparent that we orientate our lives around space, time, causes and effects, and change.

The dilemma is that the 6 days of the Genesis creation cannot be orientated with our human foundation for reality. So asking the questions when and where God created the universe, what caused the universe to be created, and what change occurred as a consequence, are completely irrelevant.
Many assume that the Genesis creation must be organized through the measurement of time, since it is divided into 7 days. The exact span of time between each day, however, is the source of debate. *The Viewpoint Theory* suggests that the 7 days represent and delineate 7 different viewpoint levels of creation.

Imagine the days of creation from a small child's perspective. When a child wakes up in the morning he doesn't think to himself that its 24 hours later than it was yesterday at this same time. Rather he perceives it much simpler than that. The child wakes up to a new day or view of creation.

> *"Truly I tell you, unless you change and become like little children, you will never enter the Kingdom of heaven."*
> *Mathew 18:3*

An eminence of God's nature is to be omnipresent (everywhere at once). From this viewpoint, reality is not divided by the measurement of time. Was the creator using a clock to create? God's creation is eternal! It is present, now. We just don't perceive it in its fullness.

A key to recognizing that God's creation and the universe we experience are not the same is to compare Genesis 1:1 with Genesis 2:4. What are the essential differences between *Genesis 1:1 "In the beginning God created the heaven and the earth,"* and Genesis 2:4, *"These are the generations of the heavens and the earth?"*

First, let's begin with Genesis 2:4. Our human foundation for reality is held within the meaning of the word *generations*. So attaching the *generations* to the heavens and the earth confines them within the domain of cause and effect, change, and space and time. How many generations of the heavens have there been? The current estimate is our universe is around 14 billion years old.

In contrast to Genesis 2:4, *Genesis 1:1 "In the beginning, God created the heaven and the earth"* reveals a reality that is created–not generated. The heaven and the earth are being created instantaneously and not through a process of change.

When we read this word "beginning" in Genesis 1:1, our first human impulse is to place it at some point in the past. We want to lock it into a moment in time by squeezing it in between events prior to it and after it. Reducing actions and events down to past moments in time, works very well for solving murder cases and remembering your birthday, but does not work for understanding Genesis 1:1.

Contemplate the immensity of this idea: If in *the beginning* God creates everything and there is nothing before it, when in time did *the beginning* in Genesis 1:1 happen? The simple truth, but frustrating to the human intellect, is that *the beginning* is not in the past. Genesis 1:1 cannot be broken down or measured as an event in time. It is an absolute act, and therefore is not related to anything else. So there are no events in time to place it within or to reference it to. *The beginning* in Genesis 1:1 is always here and now. Perhaps, this is why it is called *the* beginning and not *a* beginning.

> *"The boundaries of 'our universe' are not the boundaries of 'the universe.'"*
> –Ervin Lazlo

The word "*created*" in Genesis 1:1 is directly and deliberately correlated with the word, *generations,* in Genesis 2:4. By contrasting these two words, it will give us more insight into the differences between God's 6 days of creation and the 7th day.

If you think about the word *generations*, your family tree and ancestors may be the first things that come to mind. You might picture your father's generation, his father's generation and his father's generation and so on, going back in time. You may even look at old photographs and say, "I am a spitting image of my grandmother or great grandfather." If you dig even deeper, you may find out that your great, great grandmother was a musician, like yourself.

Explore your lineage long enough, and it may suddenly dawn on you that if your father and mother, or their father and mother, and so on, had never met; you would not be alive in your present form. So, from the standpoint of our physical existence, we embody the impressions, memories, images, experiences, and relationships of the generations which came before us. Everything that our human experience personifies is *generated* or caused by our human ancestors. This principle would also be true for all species of plants and animals.

> *"The idea is that there is a kind of memory in nature. Each kind of thing has a collective memory. So, take a squirrel living in New York now. That squirrel is being influenced by all past squirrels."*
>
> –Rupert Sheldrake.

In contrast, let's look at the word "*created*" as it pertains to Genesis 1:1. "In the beginning, God *created*..." describes an absolute timeless act. It is not a reality which is relative to the past. It is not generated by an external source. It is the creator bringing reality into existence through no other means than his omnipotent (all powerful), omniscient (all knowing), and omnipresent nature.

A very commonly posed question from thinkers and skeptics is, "If God created the universe, then who or what created God?" This question forms an impasse for logic. However, this very question can only arise from a viewpoint within the "generations," which perceives all reality as caused from past events and natural external forces. But from the viewpoint of an unlimited creator, all reality originates from within, instantaneously.

Rev 21:5 "Behold I make all things anew." If God makes all things anew, then the ultimate truth about reality is everything is now.

> *"Ultimately, all moments are one, therefore now is an eternity"*
> –David Bohm

Whenever I pass a Mexican restaurant that advertises its food as authentic, I think it must be good. God's creation unveiled in *Genesis 1:1 "In the beginning God created the heaven and the earth"* is authentic, whole, eternal and present. If I were to drive by a Mexican restaurant with a sign that said its food was an imitation, I would think it was not as good as authentic. The natural world revealed in *Genesis 2:4 "These are the generations of the heavens and the earth"* is the time bound imitation, impression, and generation of the creation. But, there is an unseen connection between the authentic creation and the natural world.

The natural world is like a memory of God's creation. Recall an event that happened in your life that you remember vividly. Now recognize that your memories, impressions, experiences, and images are *generated* from this event but are not the actual event. This is the relationship between the authentic creation and the natural world.

The verse, *Genesis 2:4 "These are the generations of the heavens and the earth, when they were created, in the day the lord God made the earth and the heavens,"* is conveying that it holds the memories, impressions, images, or generations that are within Gen 1:1. But, *Gen 1:1 "In the beginning God created the heaven and the earth"* is the unseen authentic creation!

6

The Viewpoint Theory

The Unseen Source

Hebrews 11:3 "By faith we understand that the universe was created by the word of God, so that what is seen was not made out of things that are visible."

"Everything we call real is made of things that cannot be regarded as real."
—Niels Bohr

Francis Bacon once said "knowledge is power". If the knowledge of everything was within the grasp of the intellect, that would be supreme power! So the notion of an unknowable foundation to the world can be very deflating to our ego. But, the good news is that our ability to feel extends beyond what we know.

> *"All of our reasoning ends in surrender to feeling."*
> —Blaise Pascal

Society puts knowledge at a premium. Children go to school to attain knowledge. Scientists are continually learning new things–things that were once considered unknown. So when we think of the unknown, we tend to see it in terms of something that we don't know yet, but we can know. Is there more to reality and truth than what we can know through our intellect?

> *"The true mystic is always both humble and compassionate, for she knows that she does not know."*
> —Richard Rohr

In physics, there has been a knowledge established through physical laws. The laws of thermodynamics, motion, gravity, relativity and so on define the limitations of the physical universe. It can be very comforting to know the parameters of reality. *The*

Viewpoint Theory suggests, however, that the ultimate truth of reality is limitless. The so-called physical laws of the universe are not absolute. They are natural tools that work to creatively reproduce or replicate God's unlimited creation. Therefore, viewing the universe divided by the limits of physical laws does not perceive the whole truth.

> *"Science itself is demanding a new, non-fragmentary world view."*
> –David Bohm

Most theories on Genesis believe God created and designed the universe we know. The creator shaped and set into motion substance, matter, and the physical laws of the universe long ago. A very popular theory, called Young Earth Creationism, assumes these acts took place about 6,000 years ago, over 7 literal days.

Young Earth Creationism may interpret the Genesis creation in this basic fashion: in the first 24–hour day, God creates the physical universe and light. Then on the second 24–hour day, God forms the universe. On the 3rd day, he forms the physical earth. The 4th day, he makes the physical sun and the stars. On the 5th day, he creates the birds and whales, and finally on the 6th 24–hour day, he makes the land animals, man and woman. Other theories, such as the Day–Age Theory, advocate this same premise but simply reason the 7 days as encompassing a much longer span of time.

Scientists estimate the universe began around 14 billion years ago. They believe all life was caused by biological processes over millions of years. Exhaustive studies suggest there were countless events, perhaps guided by natural selection and evolution, which caused life to begin on earth. Subsequently, it took innumerable more events to cause the first humans to show up. Finally, to get to our individual body, it required a lot of crazy coincidences.

If you contemplated all the things that had to happen for you to be here, it would seem impossible. The unplanned universe is perceived to be the source of your existence. So somehow caused by countless lucky interactions over millions and billions of years, you miraculously appeared.

God has always been the creator. Therefore, God's creation has always been. We were created in an absolute reality but we experience ourselves in a relative reality.

> *Jeremiah 1:5 "Before I formed you in the womb I knew you..."*

The Viewpoint Theory advocates that God creating the universe is an inside–out job. First is the existence of God (Being, spirit, light, and consciousness). Through this inner light and consciousness, the essence and spirit of all things are created. Then, it is through the essence and spirit and energy of all things that we perceive this world.

> *"Everything in the universe was first created in spirit and then recreated in the physical."*
>
> *–Betty Eadie NDE*

There is a difference between the existence, energy and potential of the universe and the actual experience of it. It requires a spirit viewpoint to witness God's creation of the heaven and the earth, but it requires a physical perspective—a brain to realize or experience the generations of the heavens and the earth. Our material universe exists in potential, but is not a reality unless there is life and physical consciousness.

> *"Without consciousness, "matter" dwells in an undetermined state of probability. Any universe that could have preceded consciousness only existed in a probability state."*
>
> *–Robert Lanza M.D*

Our human mind is a master at division. It invents numerous ways to separate us from nature and each other. We divide ourselves from nature through form, definitions, species, and categories and from each other through religion, race, gender, age, income, status and on and on. Our minds also divide us from the universe through space, time, and dimensions. The divisions between nature, each other, space, and time exist as a potential of God's creation, but without life and a brain, are not a reality. All the divisions we perceive create the appearance that there is an objective or independent universe out there separate from our minds.

> *"All matter originates and exists only by virtue of a force... We must assume behind this force the existence of a conscious and intelligent mind. This mind is the matrix of all matter."*
>
> *–Max Planck*

There is an unbreakable connection revealed in Genesis between spirit, consciousness and mind with the creation. In Genesis 1:1 *"In the beginning, God created the heaven and the earth,"* it was expressed that God creates everything. Yet the very next verse proclaims, Genesis 1:2, *"and the earth was without form and void and darkness was upon the face of the deep."* Genesis 1:1 states that the heavens and earth exist but Genesis 1:2 establishes they are not a reality. So how can the heaven and earth be created, yet also be void and in complete darkness?

The act of creating is not the same as the act of observing or connecting. Genesis 1:1 describes God creating everything, but does not reveal him connecting with it. The creation must be perceived and united with the creator for it to be a reality. The Divine Order in Chapter 9 will show why this is true. It is the creator's bond with the creation that brings order, illuminates and animates it. Genesis 1:2 is describing that the heaven and the earth, from that viewpoint, are disconnected from their source or creator, and thus in darkness.

The origination of darkness, described in Genesis 1:2, is the necessary gap between God creating everything and God connecting with his creation. Darkness is often equated with evil. Darkness in and of itself is not evil. The balance between darkness and light is needed for perception and experience.

> *"Only in the darkness can you see the stars."*
>
> *–Martin Luther King Jr.*

It takes a "being" to have evil. Evil is "being" worshipping the nature of darkness and destruction or rejecting God or the light. This perhaps is why Satan is often called the prince of darkness.

> *John 3:19 "The light has come into the world, and men loved the darkness rather than the light, for their deeds were evil."*

The Genesis creation uses this verse, *"And God saw that it was Good,"* throughout the 6 days. There is a very important reason for this. It is to demonstrate that the inner nature of God, which is Good, is unified with the outer creation. So it is the observer's relationship to the observed which brings the creation to life.

The Viewpoint Theory suggests that our universe is a creative imitation of God's authentic creation. Therefore, the universe's nature cannot be explained separate from us. Just like God's creation is made real through him, our universe is made real through us. Our consciousness is intertwined with the universe. It is integral with everything we observe, measure, and extrapolate.

> *"I regard consciousness as fundamental. I regard matter as a derivative from consciousness. We cannot get behind consciousness."*
>
> —Max Planck

Spirit, consciousness and perception are inner foundations of God's creation. So every little thing that we perceive like watching a sunset is interconnected to all the generations of the heavens and the earth. We have been taught that without the suns formation billions of years ago we could not enjoy the experience of a sunset. But, without conscious perception, would the sun's formation 4.5 billion years ago be a reality?

> *"Time is not an absolute reality but an aspect of our consciousness."*
>
> –Robert Lanza M.D

All conscious life is entangled with every event in linear time. So the notion that life began from non–life or consciousness originates in the brain is ultimately untrue. The late comedian, George Carlin, would often joke about people's belief in God. He would compare it to believing in the invisible man. The ironic thing is, though, that we ourselves, our very essence and consciousness, is invisible.

> *"Looking for consciousness in the brain is like looking in the radio for the announcer."*
>
> –Nasseim Harameim

If all this is mystifying, just recognize when you read the first six days of Genesis, everything that is created and made is spiritual, whole, perfect, and true. It is the spirit, essence, and potential of our physical body, material universe, linear time and space.

"...The physical body that we live in actually gives us the most beautiful vehicle ever imagined to experience time and space, and to experience the universe in ways that cannot be experienced while you are in spirit."
　　　　　　　　　　　–Mellen Thomas Benedict
　　　　　　　　　　　　Near Death Experiencer

7

The Viewpoint Theory

The Three Fundamental Principles

Psalms 46:10 "Be still and know that I am God..."

Religious movies and books have portrayed God as living in outer space somewhere, seated on a throne. God's image has been personified as an old wise male figure, divorced and detached from his creation, speaking with a very deep voice commanding energy and matter. The Genesis creation story is about the interconnection between God's nature and power with his creation. So there is a sacred and infinitely close relationship between God, creation, and us.

> *"Closer is He than breathing.*
> *And nearer than hands and feet."*
>
> *–Lord Alfred Tennyson*

The Genesis creation is organized through 3 primary fundamental principles.

1. The first principle is focused on the existence and nature of God.
2. The second principle is focused on the creation.
3. The third principle is focused on the universe and human existence.

The relationship within and between these 3 principles unveils how the Genesis creation is exquisitely organized.

The 1st Fundamental Principle:
The Existence and Nature of God

All of us can feel that we have an inner self which is independent from the outside world. We sense a division, yet simultaneously a connection, between us and creation. It is a feeling that we each have an "I" or a soul which exists in a different reality than the outer world. It may be very difficult to distinguish the inner reality of the soul from the outer reality of our body and world, but we can all feel it.

God is the great "I AM," which is free from and greater than all creation. The nature of "I AM" is revealed on the 1st day of Genesis.

The 2nd Fundamental Principle:
The Creation

When we look within ourselves, we don't usually get the impression that it is the source of all we see. For instance, you probably wouldn't look at a magnificent mountain, and suddenly realize that its immensity is a reflection of the power within you. Yet, this is the dynamic at work within God's creation. God's outer creation, revealed in the days 2 through 6, reflects the power of his inner nature. God's nature or spirit, however, is not dependent upon anything created, but the creation is powerless, dead, or in darkness when it is disconnected from "I AM" or his spirit.

The 3rd Fundamental Principle:
The 7th Day (Physical Universe and Physical Existence)

The universe began revealing to us its amazing story, after the advent of the telescope, mainly the

Hubble telescope. Like the classic fisherman story, where the fish always seems to get bigger, the universe's story is growing.

Does Genesis unveil the whole story of creation? It has been assumed that it is only describing the universe that is visible to us. *The Viewpoint Theory* proposes that it is telling the complete story. God's nature, his creation and the universe are all revealed in it. The 7th day is only a part of the story. It is the part which we are all characters in. It is unfinished from our viewpoint.

> *Ecclesiastes 3:11 "Yet God has made everything beautiful for its own time. He has planted eternity in the human heart, but even so, people cannot see the whole scope of God's work from beginning to end."*

Galileo began advocating that the sun, not the earth, was the center of our solar system in the early 1600s. Since then, science has been knocking us humans down on the scale of our significance. It has exposed the humbling fact that the imprint of our entire human history is very small compared to the giant universe. Ultimately, though, it is God's present and unseen creation which sustains the entire universe.

We literally live within the 7th day, or the generations of the heavens and the earth. So our natural inclination is to place God's 6 days of creation in the past. Thus, we move it far away from the presence of God. God and his creation are eternally linked in the present. Therefore, they are undividable by space and time. They are always near to us.

> *Mathew 3:2 "Repent for the kingdom of heaven is at hand."*

Most believe that God's 6 days of creation is somewhere, in the distant past. So if we had time machine, we could travel back in time and watch God creating the universe. The universe that we observe through laws, linear time, and forces is contained in the 7th day. This world is not God's finished creation. Its potential, however, is held within it.

> *Mathew 13:31 "The Kingdom of heaven is like a mustard seed, which a man took and planted in his field..."*

The days 2 through 6 are like a tree seed. (this is the message in Genesis 2:5–2:7). The seed holds the knowledge and potential of the tree within it. If the seed receives the right mixture of nutrients, sunlight, rain, and so forth, it becomes a tree. God's creation (6 days) is like a seed. It holds the information of (7th day) universes, time, space, galaxies, stars, planets, and life within it.

> *"The information within a system is more fundamental than matter, energy, space and time. Information is the one thing in the universe that cannot be destroyed."*
>
> —Ervin Laszlo

Imagine that you were witnessing the ocean's vastness for the very first time. The instance you see its enormity and power, a response, impression, emotion or experience would be invoked within you. The origination of our reality could be described as a response within us to the power of God's unseen creation.

> *"All I have seen teaches me to trust the creator for all I have not seen."*
>
> —Ralph Waldo Emerson

When God "rests" on the 7th day, it allowed us to receive, interpret, and respond to him, his intentions and creation. The universe and our reality is the evidence of how we are receiving God's creation. But, even as magnificent as this universe is, the potential in what God and his creation can offer is far greater than how we are experiencing it. The universe, sun, stars, and colors we see, the sounds and music we hear, and the joy and love we feel are only a fragment of their potential.

"Those that have had near death experiences will tell you that realm is far more real than this world, more crisp, vibrant, alive."

–Dr. Eben Alexander

We tend to interpret God's creation through the limitation of our human brain, which has been conditioned by our personal past and all past generations. This confines our experiences within a universe of limited perception and laws. So we don't see the limitlessness, beauty, goodness, timelessness and wholeness of God's creation. Those people that have died and then came back to life seem to describe experiencing the creation closer to its fullness.

John 3:3 "...No one can see the Kingdom of God unless they are born again."

8

The Viewpoint Theory

Writing the Genesis Creation

> *"Intuition enlightens and so links up with pure thought. They together become an intelligence, which is not simply of the brain, which does not calculate, but feels and thinks."*
>
> –Piet Mondrian

For most of us, we intuitively feel that God exists, even though we may not have a logical reason why. God is literally beyond reasoning! The *Viewpoint Theory* was born from faith and a feeling–an intuition that there is much more to reality than the intellect can comprehend. But from that faith, feeling and intuition a divine order emerged–an order that you can reason.

The Viewpoint Theory began as a sequence of insights about the creation of the universe. Several years after its inception, it became an interpretation of the Genesis creation story. The commonality is that both are based on the same divine order (Chapter 9). To understand The Divine Order, you need imagination, intuition and reasoning. Imagination and intuition may seem to contradict reasoning. But when they have been synchronized together, it has led to some of the greatest inventions and discoveries in human history. Let your imagination and intuition guide your reasoning!

> *"Intuition will tell the thinking mind where to look to next."*
> –Jonas Salk

There is a reason why the Genesis creation is organized exactly the way it is. The author did not simply make it all up. He didn't just walk outside one day, look around and create a fairytale about how God made light, stars, earth, humans and animals in 7 days. There is a profound reasoning at the core of it. It is not a reasoning based on any kind of scientific laws, theories, and principles. It is not an inductive, deductive, or abstract reasoning used to solve math problems, build puzzles, see patterns, and understand symbols. It is reasoning based in the connection between the nature and viewpoint of the observer with the observed. Beginning from an unlimited creator and observer, there is a rational process to get to our universe and us.

> *"That deep emotional conviction of the presence of a superior reasoning power, which is revealed in the incomprehensible universe, forms my idea of God."*
> *–Albert Einstein*

Even though our intellect cannot grasp God, there is an eternal relation between God and us. So there is a path and foundation which connects the highest viewpoint of God to me looking out over the ocean. I call this timeless path and foundation The Divine Order.

Imagine and feel beyond all natural laws and the visible world. Use your imagination and intuition to guide you into viewpoints which transcend science. There is a knowing within all of us that there is an invisible foundation which allows the universe to be visible. From this intuitive knowingness, you can use reasoning to discover The Divine Order. It's like a feeling or knowing without seeing, that there is a mountain over the horizon with a view from on top of it. You then can deduce that from that view you can see the whole city.

God has a view of creation which we cannot see, yet we have faith it is there. You start from the highest viewpoint, and then climb down the mountain or levels of creation until you get to your personal viewpoint. Genesis begins from the highest viewpoint (the whole) and ends with our viewpoint (the part).

God's 6 days in Genesis describe a vertical account of creation. Nothing that is created is bound by the confines of the past. The 7th day is a linear time description of creation. Everything described lies within the boundaries of what has been.

There are twelve divine acts within the 7 days of the Genesis creation story. In the next chapter, the order and purpose behind each of the 12 divine acts will be revealed.

In the following pages, I will present two creation narratives. First, will be the Genesis creation story. Then, will be my version. You can contrast and compare the two stories through the 12 divine acts. In my version, I apply *The Viewpoint Theory* to intuitively reason how an unlimited creator would create the universe. I borrow some of the same vocabulary words used in Genesis to illustrate the correlation, but the order and most of the words are simply natural. Am I using the same intuitive reasoning as the Genesis creation story? You decide!

The location of the **12** divine acts within the Genesis creation is shown below. There is a sacred connection between the first **6** divine acts and the second **6** divine acts, which will be revealed in the next chapter.

The Genesis Creation Story:
Genesis 1 and 2
(King James Version of the Bible)

1st Divine Act
1 In the Beginning God created the heaven and the earth.

2 And the earth was without form, and void; and darkness was upon the face of the deep.

2nd Divine Act
And the spirit of God moved upon the face of the waters.

3rd Divine Act
3 And God said, let there be light: and there was light.

4th Divine Act
4 And God saw the light, that it was good:

5th Divine Act
And God divided the light from the darkness.

6th Divine Act
5 And God called the light day, and the darkness he called night. And the evening and the morning were the **first day.**

7th Divine Act
6 And God said, let there be a firmament in the midst of the waters, and let it divide the waters from the waters.
7 And God made the firmament, and divided the waters which were under the firmament from the waters which were above the firmament: and it was so.
8 And God called the firmament heaven. And the evening and the morning were the **second day.**

8th Divine Act
9 And God said let the waters under the heaven be gathered together unto one place, and let the dry land appear: and it was so.
10 And God called the dry land earth; and the gathering together of the waters he called seas: and God saw that it was good.
11 And God said, let the earth bring forth grass, the herb yielding seed, and the fruit tree yielding fruit after his kind, whose seed is in itself, upon the earth: and it was so.

12 And the earth brought forth grass, and herb yielding seed after his kind, and the tree yielding fruit, whose seed was in itself, after his kind: and God saw that it was good.

13 And the evening and the morning were the **third day**.

9th Divine Act

14 And God said, let there be lights in the firmament of heaven to divide the day from the night; and let them be for signs, and for seasons, and for days, and years:

15 And let them be for lights in the firmament of heaven to give light upon the earth: and it was so.

16 And God made two great lights; the greater light to rule the day, and the lesser light to rule the night: he made the stars also.

17 And God set them in the firmament of heaven to give light upon the earth,

18 And to rule over the day and over the night, and to divide the light from the darkness: and God saw that it was good.

19 And the evening and the morning were the **fourth day.**

10th Divine Act

20 And God said, let the waters bring forth abundantly the moving creature that hath life, and fowl that may fly above the earth in the open firmament of heaven.

21 And God created great whales, and every living creature that moveth, which the waters brought forth abundantly, after their kind, and every wing fowl after his kind: and God saw it was good.

22 And God blessed them, saying, be fruitful, and multiply, and fill the waters in the seas, and let fowl multiply in the earth.

23 And the evening and the morning were the **fifth day**.

11th Divine Act

24 And God said, let the earth bring forth the living creature after his kind, cattle, and creeping thing, and beast of the earth after his kind: and it was so.

25 And God made the beast of the earth after his kind, and cattle after their kind, and everything that creepeth upon the earth after his kind: and God saw that it was good.

26 And God said, let us make man in our image, after our likeness: and let them have dominion over the fish of the sea, and the fowl of the air, and over the cattle, and over all the earth, and over every creepy thing that creepeth upon the earth.

27 So God created man in his own image, in the image of God created he him; male and female created he them.

28 And God blessed them, and said unto them, be fruitful, and multiply, and replenish the earth, and subdue it: and have dominion over the fish of the sea, and over the fowl of the air, and every living thing that moveth upon the earth.

29 And God said, behold, I have given you every herb bearing seed, which is upon the face of all the earth, and every tree, in the which is the fruit of a tree yielding seed; to you it shall be for meat: and it was so.
30 And to every beast of the earth, and to every fowl of the air, and to everything that creepeth upon the earth, wherein there is life, I have given every green herb for meat: and it was so.
31 And God saw everything that he had made, and, behold, it was very good. And the evening and the morning were the **sixth day.**

12th Divine Act
1 Thus the heavens and the earth were finished and all the host of them.
2 And on the seventh day God ended his work which he had made; and he rested on the seventh day from all his work which he had made.
3 And God blessed the seventh day, and sanctified it: because that in it he had rested from all his work which God created and made.
4 These are the generations of the heavens and the earth when they were created, in the day that the lord God made the earth and the heavens.

My Creation Narrative

1st Divine Act
God is omnipotent, omniscient, and omnipresent. Therefore, God instantaneously creates the essence

of everything. This includes what we call the heaven and the earth.

From this viewpoint, however, the creator is beyond his creation. So the creation is disconnected from its source and there is no experience of it. Therefore, the heaven lies in complete stillness, absolute darkness, and is unseen, and the earth is formless and void of energy and life.

2nd Divine Act
God connects his power source or invisible spirit to the essential substance of creation. Through the power of his spirit, he can bring energy to all things.

However, God's spirit is invisible from this viewpoint.

3rd Divine Act
God reveals his spirit through absolute light. Perfect and unlimited light naturally unveils the nature of his spirit.

From this viewpoint, however, there is no experience of the light.

4th Divine Act
God receives the impression of his spirit's light. This impression creates infinite consciousness and contains inner qualities of existence beyond our comprehension. God's impression of the light

creates a sense of being or self-awareness through which all reality can be experienced.

From this viewpoint, however, there is no perception or experience of the inner reality of being apart from or in contrast to the outer reality of creation.

5th Divine Act
God divides the inner light of spirit from the outer darkness of creation.

From this viewpoint, however, the inner light of spirit does not reveal the outer creation. The absolute and perfect light of God's spirit is incompatible with the absolute darkness wherein the creation exists.

6th Divine Act
God creates a new kind of light—a light that can reveal his creation. He creates a relative light from the absolute light of spirit. This relative light has what we might think of as positive charge to it. So that it can project into the darkness and give light to his creation. He then creates a negative charge to the absolute darkness, so that the relative darkness could receive this light. This is the 1st viewpoint of creation.

From this viewpoint of creation however, everything is one. There is no division in creation. There is no view or experience from within it.

7th Divine Act

God creates space within all the substance of creation to embody the power of his nature. This space within all creation which expresses the power of the creator we call heaven. Through heaven, the creator can connect to, see, and observe all things. This is the 2nd viewpoint of creation.

From this viewpoint of creation however, there is no form. There is no reference point within creation through which the power of God's spirit can be expressed.

8th Divine Act

God creates an energy form from the essential substance of creation. We call this form, earth. This energy form sustains the essential substances like water, elements, and minerals needed for life. So God's spirit moving through the energy form of the earth creates the seed and potential for all life. Keep recognizing that this is God's view of creation, therefore it reflects his nature. This is the 3rd viewpoint of creation.

From this viewpoint of creation however, there is no relationships to the earth which can reveal its potential. There is no form which can embody and express the power in the light of God's spirit.

9th Divine Act

God creates energy light forms which can relate to the earth form. He creates the sun, stars, and all the celestial bodies of the universe. These energy light bodies have the ability to interact with and reveal the potential of the earth. They allow us the ability to perceive time and change. They give the earth light, the potential for seasons, stability and predictability. Keep recognizing this is God's view of creation, therefore it reflects God's nature. This is the 4th viewpoint of creation.

From this viewpoint of creation, however, there is no conscious form. There is no form which can embody and express the power in the impression of spirit light. There are no forms through which to experience the relationships of the essential substances of the earth to the sun, stars and moon.

10th Divine Act

God creates conscious forms, which can move and interact with the environment. Keep recognizing this is God's view of creation. Therefore, it reflects God's nature. This is the 5th viewpoint of creation.

From this viewpoint of creation however, there is no conscious perception or mind through form. There is not a form which can embody and express spiritual perception.

11th Divine Act
God creates and divides all the life forms on earth through their ability to perceive and understand. He creates humanity with the potential to embody and express spirit perception. He endows them with the capacity to understand the world and universe. Keep recognizing that this is God's view of creation; therefore it reflects God's nature. This is the 6th viewpoint of creation.

From this viewpoint, however, there is no material universe. There is no experience of physical change through a body or form. There is no cause and effect and a past, present and future. There are no generations of the earth, heavens, life, and humanity.

12th Divine Act
God lets go of his viewpoint of creation in order for it to be experienced through linear time and a limited perspective. He allows everything he created to be manifested through a process of causes and effects and physical change. He allows the natural world to creatively imitate his authentic creation, so it can appear to evolve and function on its own through physical laws, probabilities and chance. He allows the human form to experience the world through free will. Recognize that this is a human view of creation. This is the 12th viewpoint of creation.

9

The Viewpoint Theory

The Divine Order

"I want to know how God created this world. I am not interested in this or that phenomenon, in the spectrum of this or that element. I want to know God's thoughts. The rest are details."
—Albert Einstein

The puzzle that is the Genesis creation story is taken apart and explored in this chapter. You will see how all the pieces fit together. The universe we know is but one piece of the whole puzzle. Only when you can comprehend the total picture, can you have the right perspective of science and a creator.

The 6 days in Genesis describe viewpoints that our eyes cannot look through, but our minds can imagine through. These viewpoints are based in a divine order. I call it The Divine Order because it originates from the sacredness of Being.

Physicists form elaborate theories like string theory and ideas like multi-verses to explain our reality and universe, but they have yet to find a creator and spiritual foundation for everything.

The 7 days of God's creation is the spiritual foundation of everything. It is organized through 12 sacred actions. The number 12 in the bible represents completion. There are 6 inner divine acts on the 1st day. They reveal the existence and nature of God. There are 6 outer divine acts which correspond with the days 2 through 7. I call them the 6 Outer Creative Acts. They reveal God's creation and the universe. The 6 inner divine acts correlate with the 6 outer creative acts.

Each inner divine act and outer creative act establishes the foundation for all that is. The numbers, 1 through 6, are the inner foundations of existence, and 7 through 12 are the outer foundations of creation.

The 6 inner foundations are qualities of being. They exist independent of substance and form and cannot be fully defined in any physical or measurable way.

The 6 inner foundations of existence are:
1. Creator
2. Spirit
3. Light
4. Consciousness
5. Mind (perception)
6. Time or change

The 6 inner foundations of existence (1-6) are unveiled on the 1st day in Genesis.

The 1st inner Divine act
1. *Gen 1:1 "In the beginning God created the heaven and the earth.* The 1st inner divine act (Gen 1:1) reveals that God is the **creator**. It does not reveal the outer **creation** as described in Gen 1:2. *Gen 1:2 "and the earth was without form and void and darkness was upon the face of the deep."*

The 2nd inner Divine act
2. *Gen 1:2 "And the spirit of God moved upon the face of the waters."* The 2nd inner divine act (Gen 1:2) reveals God's **spirit**. It does not reveal the outer creation of **form**.

The 3ʳᵈ inner Divine act
3. *Gen 1:3 "And God said let there be light and there was light"*. The 3ʳᵈ inner divine act (Gen 1:3) reveals **the light**. It does not reveal the outer creation of **light forms**.

The 4ᵗʰ inner Divine act
4. *Gen 1:4 "And God saw the light that it was Good."* The 4ᵗʰ inner divine act (Gen 1:4) reveals **consciousness.** It does not reveal the outer creation of **conscious forms.**

The 5ᵗʰ inner Divine act
5. *Gen 1:4 "And God divided the light from the darkness."* The 5ᵗʰ inner divine act (Gen 1:4) reveals **mind (perception).** It does not reveal the creation of living **forms with a mind.**

The 6ᵗʰ inner Divine act
6. *Gen 1:5 "And God called the light day and the darkness he called night."* The 6ᵗʰ inner divine act (Gen 1:5) reveals **time or change**. It does not reveal the outer creation of **forms changing**.

The 6 outer foundations are defined through substance and form. They are the foundations which allow us to experience divisions between each other and all things through time and space.

The 6 outer foundations are:
1. Creation (substance)

2. Form
3. Light form
4. Conscious form
5. Mind (perception) identified with form
6. Forms changing.

The 6 outer Divine acts or foundations are unveiled on the days 2 through 7 in Genesis.

The 1st outer Divine act
1. *(2nd day) Gen 1:6 "And God said, let there be a firmament in the midst of the waters, and let it divide the waters from the waters."*
The 1st outer creative act reveals God's **creation**.

The 2nd outer Divine act
2. *(3rd day) "Gen 1:9 "And God said let the waters under the heaven be gathered together unto one place, and let the dry land appear: and it was so."*
The 2nd outer creative act reveals the creation of **form**.

The 3rd outer Divine act
3. *(4th day) Gen 1:16 "And God made two great lights; the greater light to rule the day, and the lesser light to rule the night: he made the stars also."* The 3rd outer creative act reveals the creation of **light forms**.

The 4th outer Divine act
4.. (*5th day*) *Gen 1:21 "And God created great whales, and every living creature that moveth, which the waters brought forth abundantly, after their kind, and every wing fowl after his kind: and God saw it was good."* The 4th outer creative act reveals the creation of **conscious forms**.

The 5th outer Divine act
5. (*6th day*) *Gen 1:27 "So God created man in his own image, in the image of God created he him; male and female created he them."* The 5th outer creative act reveals the creation of a **mind that is identified with form or body.**

The 6th outer Divine act
6. (*7th day*) *Gen 2:2 "and on the 7th day God ended his work which he had made ; and he rested on the 7th day from his work which he had made"* *Gen 2:4 "These are the generations of the heavens and the earth…"* The 6th outer act reveals **forms changing through time.**

If you combine the 6 inner foundations of existence with the 6 outer foundations of creation, it makes a total of 12.

The 6 Inner Foundations:
1. Creator
2. Spirit
3. Light

4. Consciousness
5. Mind (perception)
6. Time or Change

The 6 Outer Foundations:
7. Creation (substance)
8. Form
9. Light Forms
10. Conscious forms
11. Mind (perception) identified with forms
12. Forms changing

The hidden code that the Genesis creation is organized through is **1-7, 2-8, 3-9, 4-10, 5-11,** and **6-12**. The 6 inner divine acts **(1-6)** parallel the 6 outer creative acts **(7-12)**. This inner-outer contact brings order, energy and life to creation and form.

In review, the 6 inner foundations of existence are revealed on the 1st day of Genesis. The 6 outer foundations of creation are revealed on the days 2 through 7. But, also revealed on the days 2 through 7 are the 6 aspects that make Gods outer creation a reality.

The power behind all reality originates from within the nature of God. So when the inner foundations of existence flows through the outer foundations of creation, the potential within everything God created is realized.

The potential of God's creation is realized through:

1. Heaven *(2nd day)*
2. Life *(3rd day)*
3. Energy (relationships) *(4th day)*
4. Living bodies *(5th day)*
5. Memory *(6th day)*
6. Generations *(7th day)*

1. **Creator/creation-** *(2nd day)* Gen 1:7 *"And God made the firmament, and divided the waters which were under the firmament from the waters which were above the firmament: and it was so. Gen 1: 8 and God called the firmament heaven."* The power in the **(1) creator** connecting to **(7) creation** is embodied in heaven. **Heaven** brings order and reality to all creation.

2. **Spirit/Form.** *(3rd day)* Gen *1:10 "And God called the dry land earth; and the gathering together of the waters he called seas: and God saw that it was good." 11 And God said, let the earth bring forth grass, the herb yielding seed, and the fruit tree yielding fruit after his kind, whose seed is in itself, upon the earth: and it was so."* The power in the creator connecting **(2) spirit** with **(8) form** creates the potential for all **life.**

3. **Light/form.** *(4th day)* Gen 1:14 *"And God said, let there be lights in the firmament of heaven to divide the day from the night; and let them be for*

signs, and for seasons, and for days, and years: 15 And let them be for lights in the firmament of heaven to give light upon the earth: and it was so" ... 17 And God set them in the firmament of heaven to give light upon the earth, 18 And to rule over the day and over the night, and to divide the light from the darkness: and God saw that it was good. The power in the creator connecting **(3) light** with **(9) form** creates the potential for **energy and relationships**. It causes interaction between forms, particles, chemicals and so on. It generates forces, patterns, tendencies, and any other observable phenomenon.

4. **Consciousness/Form** *(5th day) Gen 1:20 "And God said, let the waters bring forth abundantly the moving creature that hath life, and fowl that may fly above the earth in the open firmament of heaven."* The power in the creator connecting **(4) consciousness** with **(10) form** creates the potential for **living bodies.** It creates the ability for living bodies to move and interact with the environment and natural laws.

5. **Mind (perception)/Form** *(6th day) Gen 1: 26 "And God said, let us make man in our image, after our likeness: and let them have dominion over the fish of the sea, and the fowl of the air, and over the cattle, and over every creepy thing that creepeth upon the earth."* The power in the creator connecting **(5) mind (perception)** with **(11) form** creates the potential for bodies to hold certain

memories. Every species and living animal has within its body the memory of what it is.

6. **Change/Form.** *(7th day) Gen 2:4 "These are the generations of the heavens and the earth when they were created in the day that the lord God made the earth and the heavens". Genesis 2:5–2:7 "And every plant in the field before it was in the earth, and every herb of the field before it grew, for the lord God had not caused it to rain upon the earth and there was not a man to till the ground. But there went up a mist from the earth, and watered the whole face of the ground. And the lord God formed man of the dust of the ground, and breathed into his nostrils the breath of life; and man became a living soul."* The power in the creator allowing **(6) time or change** to connect with **(12) form** creates the potential for the "big bang," **generations**, and a changing and evolving universe. It allows for everything that God created to be creatively reproduced through cause and effect and natural processes.

The following table presents The Divine Order. It shows the interconnections between the 6 inner foundations of existence and the 6 outer foundations of creation. It also reveals the aspects that allow the creations potential to be realized.

The Divine Order

1st Day of Genesis The 6 Inner Foundations of Existence	Days 2–7 of Genesis The 6 Outer Foundations of Creation
1. (*Gen. 1:1*) Creator	**7.** (*2nd Day*) Creation—Heaven
2. (*Gen. 1:2*) Spirit	**8.** (*3rd Day*) Form (Earth)—Life
3. (*Gen. 1:3*) Light	**9.** (*4th Day*) Form—Energy
4. (*Gen. 1:4*) Consciousness	**10.** (*5th Day*) Form—Living Bodies
5. (*Gen. 1:4*) Mind (Perception)	**11.** (*6th Day*) Form—Memory
6. (*Gen. 1:5*) Time or Change	**12.** (*7th Day*) Form—Generations

We are interconnected with the 6 days of creation. Our being and potential originates through the power of God's nature.

> *Acts 17:28 "For in him we live, and move, and have our being."*

Most theories in science try to explain creation and existence separate from a creator and us. Any scientific theory on the universe's origin which does not include God and consciousness will be incomplete. Let's explore deeper the dynamic between God, creation and us in the Genesis creation story.

Creator (Gen 1:1)

Genesis 1:1 "In the beginning, God created the heaven and the earth" is the 1st divine act, revealing the existence and nature of God. Recognize that you are a creator. You have the power to create.

> *Mathew 19:26 "...With God all things are possible."*

Genesis 1:1 clearly proclaims that God exists and that he is the omnipotent creator of everything. It does not, however, unveil the creation, as indicated in Genesis 1:2 *"And the earth was without form and void and darkness was upon the face of the deep."* It only acknowledges that God is the creator and he created the heaven and the earth instantaneously or in the beginning. On the 2nd day of Genesis, God divides the substance (water) of creation with "heaven," uncovering his creation.

Creation (2nd day)

There is a sacred connection between the creator (Gen 1:1) and creation (2nd day). The power in this union creates the potential for the visible universe.

6 And God said, let there be a firmament in the midst of the waters, and let it divide the waters from the waters.
7 And God made the firmament, and divided the waters which were under the firmament from the waters which were above the firmament: and it was so.
8 And God called the firmament heaven. And the evening and the morning were the second day.

The 2nd day is about God establishing an interaction and relationship between his nature and creation through heaven.

Genesis 1:1 reveals the omnipotent creator, and the 2nd day unveils God's creation through Heaven. Heaven is where the power of the creator's nature is reflected, expressed, and embodied in the creation, "Our father who art in heaven....."

We relate to heaven through the sky and space. But, what we are witnessing is the generations or impressions of the heavens. God's creation of heaven is much more than the cosmos.

> *"The heaven I experienced was so pure, loved filled, and magnificent that I did not want to return to earth."*
> *–Mary C. Neal, M.D*
> *Near Death Experiencer*

You can relate to the 2nd day on a personal level. Recognize that you have the power to create, yet you are connected to all of God's creation through heaven.

Spirit (Gen 1:2)

Genesis 1:2 "And the spirit of God moved upon the face of the waters" is the 2nd divine act, revealing the existence and nature of God. Spirit is God's formless inner connection to creation.

John 4:24 "For God is spirit, so those who worship him must worship in spirit and in truth."

God's spirit can move beyond the substances of creation. The spirit is *of* God not *of* creation. It is supernatural, and therefore not bound by the laws of the natural world. Recognize that you are spirit, free from all physical forces and laws.

"Our eternal spiritual self is more real than anything we perceive in this physical realm and has a divine connection to the infinite love of the creator."

–Eben Alexander

Genesis 1:2 demonstrates that spirit is God's inner relationship to his creation. However, it does not unveil an outer connection or reference for creation. This is revealed on the 3rd day of Genesis.

Form (Earth) (3rd day)

There is a sacred connection between spirit (Gen 1:2) and form (water-earth) 3rd day, which creates the potential for all life.

9 And God said let the waters under the heaven be gathered together unto one place, and let the dry land appear: and it was so.
10 And God called the dry land earth; and the gathering together of the waters he called seas: and God saw that it was good.
11 And God said, let the earth bring forth grass, the herb yielding seed, and the fruit tree yielding fruit after his kind, whose seed is in itself, upon the earth: and it was so.
12 And the earth brought forth grass, and herb yielding seed after his kind, and the tree yielding

fruit, whose seed was in itself, after his kind: and God saw that it was good.
13 And the evening and the morning were the third day.

The 3rd day is about the creator establishing an interaction and relationship between the spirit and water-earth through life.

We perceive the earth as this physical planet. But what we are witnessing is the generations or impressions of the earth. We don't see God's finished creation of earth.

> *Job 38:4 "Where were you when I created the earth?"*

On the 3rd day the creator forms an outer reference point for reality, called earth. The earth serves as a home within creation to express, reflect and embody the power of God's spirit. The power of spirit is demonstrated through all the varieties of life on earth.

You can relate to the 3rd day. Recognize that your spirit is free from form, yet it has the gift to connect to or be form.

Light (Gen 1:3)

Genesis 1:3 "And God said let there be light: and there was light" is the 3rd divine act, revealing the existence and nature of God. The light in Genesis 1:3 is absolute, perfect, and immeasurable. It openly expresses the nature of God's spirit.

John 1:5 "...God is light; in him there is no darkness at all."

The light source in Genesis 1:3 originates directly from God and through his spirit. It is eternally sustained from within. It is miraculous, without limitation, and needs no external force for it to illuminate. So no batteries required. It is not comparable to the light that we see.

Revelation 21:23 "The city does not need the sun or the moon to shine on it, for the glory of God gives it light, and the lamb is its lamp".

You can relate to Genesis 1:3 on a personal level. Recognize that you emanate a light from within your spirit that is untouched by anything in the world.

1 John 4:4 "... greater is he that is in you, than he that is in the world."

The light in Genesis 1:3 illuminates the inner nature of God's spirit. However, it does not shine on the earth form. The earth form is revealed on the 4th day of Genesis through the light emanating from celestial bodies.

Light Form (4th day)

There is a sacred relationship between light (Gen 1:3) and Form (4th day), which creates the potential for all energy, relationships and observations.

14 And God said, let there be lights in the firmament of heaven to divide the day from the night; and let them be for signs, and for seasons, and for days, and years: 15 And let them be for lights in the firmament of heaven to give light upon the earth: and it was so.
16 And God made two great lights; the greater light to rule the day, and the lesser light to rule the night: he made the stars also.
17 And God set them in the firmament of heaven to give light upon the earth,
18 And to rule over the day and over the night, and to divide the light from the darkness: and God saw that it was good.
19 And the evening and the morning were the fourth day.

The 4th day is about the creator establishing an interaction and relationship between heaven and earth through light or energy forms.

The power of the absolute light unveiled in Genesis 1:3 is reflected, embodied, and expressed through celestial forms on the 4th day. The potential for all physical processes, calculations, measurements, relationships, laws or tendencies and forces originate from it. The earth's atmosphere, environment, rotation, seasons, orbit, and ocean tides are effects of the interaction between heaven and earth.

We perceive the 4th day as the material sun and stars, which are made of gases and elements. But what we are witnessing is the generations or reproductions of God's creation.

We can relate to the 4th day on a personal level. Recognize that the light of your spirit is free from form, yet it has the power to emanate through and as form.

Consciousness (Gen 1:4)

Genesis 1:4 "And God saw the light that it was good" is the 4th divine act, revealing the existence and nature of God. Genesis 1:4 demonstrates that God's impression of his spirit's light is that it was Good.

> *"...A tremendous amount of warmth and love came from the light..."*
> –Robin Michelle Halberdier
> *Near Death Experiencer*

The Viewpoint Theory defines consciousness as the impression of being, that is independent from form, substance, or body through which reality is experienced. The word *Good* in Genesis 1:4 contains truth, perfection, unlimited and love. So the impression of the light is that it is true, love, unlimited and all other attributes of *Good*. The light in essence is life.

> *John 14:6 "I am the way and the truth and the life. No one comes to the father except through me."*

Recognize that you have an impression of being or sense of self, called consciousness. Through this inner light of consciousness, you can experience reality beyond your physical body and brain.

Genesis 1:4 illuminates God's impression of the light. It does not, however, identify the light within living forms. This is revealed on the 5th day. The power in the impression of the light is embodied, reflected, and expressed through conscious forms.

Conscious Form (5th day)

There is a sacred connection between consciousness (Gen 1:4) and form (5th day), which creates the potential for all conscious life forms.

20 And God said, let the waters bring forth abundantly the moving creature that hath life, and fowl that may fly above the earth in the open firmament of heaven.
21 And God created great whales, and every living creature that moveth, which the waters brought forth abundantly, after their kind, and every wing fowl after his kind: and God saw it was good.
22 And God blessed them, saying, be fruitful, and multiply, and fill the waters in the seas, and let fowl multiply in the earth.
23 And the evening and the morning were the fifth day.

The 5th day is about the creator establishing an interaction and relationship between consciousness with the atmosphere, sky, water-seas, and environment through moving life forms. It describes the movement of whales through the seas and birds through the heaven or sky.

The Viewpoint Theory defines conscious form as an impression of being that is identified with form through which reality is experienced.

We think of the creatures, whales and birds, which God creates on the 5th day, as the ancestors of the life forms we see today. But, all ancestors of the sea creatures and birds alive today are also the generations of God's creation.

You can relate to the 5th day on a personal level. Recognize that the impression of your being, or sense of self, is free from your form, or body. Yet, it has the power to be identified with a form or body.

Our consciousness comprises three kinds of awareness. First, at our core we have a sense of being that is free from all form and limits. Second, we have a sense of being that is identified with a spiritual or energy form. Third; we have a sense of being that is identified with a physical form, brain and a human body. The 5th day describes consciousness identified with a spiritual or energy form.

> *"...None of us are as "real" or as physical as we think we are. From what I saw it looked like we are energy first, and physical is only a result of expressing our energy..."*
> –Anita Moorjani
> Near Death Experiencer

Mind (Perception) (Gen 1:4)

Genesis 1:4 "And God divided the light from the darkness" is the 5th inner divine act, revealing the existence and nature of God. It unveils the division between the inner light and the outer darkness.

The mind is what allows us to perceive a division between inner reality and outer reality.

It gives us the ability to sense that we are not creation. Recognize that you have a mind which extends beyond your body.

Genesis 1:4 is the act of dividing the light of being from the outer darkness of creation. It does not, however, reveal the ability for living forms to perceive themselves as divided from other forms and the environment. That awareness is revealed on the 6th day. The light of being divided from, or in contrast to, the outer creation is embodied, reflected, and expressed in the 6th day through image or form.

Mind (Perception) through Forms (6th day)

There is a sacred connection between mind (perception) Gen 1:4 and form (6th day), which creates the potential for conscious living forms to perceive and interact with the world around them.

24 And God said, let the earth bring forth the living creature after his kind, cattle, and creeping thing, and beast of the earth after his kind: and it was so.
25 And God made the beast of the earth after his kind, and cattle after their kind, and everything that creepeth upon the earth after his kind: and God saw that it was good.
26 And God said, let us make man in our image, after our likeness: and let them have dominion over the fish of the sea, and the fowl of the air, and over

the cattle, and over all the earth, and over every creepy thing that creepeth upon the earth.
27 So God created man in his own image, in the image of God created he him; male and female created he them.
28 And God blessed them, and said unto them, be fruitful, and multiply, and replenish the earth, and subdue it: and have dominion over the fish of the sea, and over the fowl of the air, and every living thing that moveth upon the earth.
29 And God said, behold, I have given you every herb bearing seed, which is upon the face of all the earth, and every tree, in the which is the fruit of a tree yielding seed; to you it shall be for meat: and it was so.
30 And to every beast of the earth, and to every fowl of the air, and to everything that creepeth upon the earth, wherein there is life, I have given every green herb for meat: and it was so.
31 And God saw everything that he had made, and, behold, it was very good. And the evening and the morning were the sixth day.

The 6th day is about the creator establishing an interaction and relationship between mind or perception with heaven, earth, nature and environment through animals and the human life form.

You can relate to the 6th day on a personal level. Recognize that you have a mind which extends beyond the body that divides you from creation. Yet, you also have a human brain which divides your perceptions and memories and allows you to experience other life forms and nature.

There is mystery that surrounds the 6th day. Genesis 1:26. *"And God said let us make man in our own image and our likeness."* The reason why this verse refers to the absolute singular God in a relative plural sense is to demonstrate that Being or spirit is identifying itself with a form or image. Therefore, it can experience itself as separate from other images or energy forms.

Time (Gen 1:5)

Genesis 1:5 "And God called the light day and the darkness he called night" is the 6th inner divine act revealing the existence and nature of God. Genesis 1:5 demonstrates that through the inner light of spirit, God creates time or change by calling the light *day* and darkness *night*. Day and night is a metaphor for time. Time is the medium that connects inner being with outer reality.

In Genesis 1:4, "And God divided the light from the darkness," it was made clear that the absolute light of spirit was incompatible with the absolute darkness of creation. Genesis 1:4 revealed that there was no relationship between the light and creation. So by calling the light "day," God creates an impression of the light that can illuminate his creation.

The light in Gen 1:3 is absolute and the light God calls *day* is relative. So recognize that the light called *day* and the light God creates in Gen 1:3 are, in essence, two different lights. For instance, if you were to look at an art painting and call it either pretty or ugly, your impressions would exist as a separate reality from the actual painting. So God calling the light *day* creates a reality within or apart from the absolute light in Gen 1:3.

Here is another analogy to perhaps bring clarity to Genesis 1:5. Visualize the word *day* as an impression of the light that possesses a positive charge. God calls the darkness *night*, so it can receive the light of day. So envision the word *night* as an impression of the darkness that has a negative charge. *Day* has a positive charge to project light and *night* has a negative charge to receive light. Behind the *day and the night,* the absolute light and darkness still exists.

Absolute light cannot reveal what is concealed in absolute darkness, but the relative day can shine into what is hidden in the relative night. This is the meaning of Genesis 1:5. So in the morning at the end of the first day, God's creation comes into the light of day. On the 2nd day is the 1st outer creative act, revealing God's creation.

Time can be a hard concept to grasp because we think of it as outside of us. However, Genesis 1:5 shows that time is created by a shift in perception from inner reality to an outer reality. So the light of time moves, or is projected from, the power of inner existence to an outside creation.

> *Revelation 22:13 "I am the alpha and the omega, the first and the last, and the beginning and the end."*

Free Will

We have the ability within us to create change. Recognize that you have the power to shift perception or decide your actions, which is free will. Free will is what allows your soul to engage with the outer world. It is important to realize this distinction, though. Free will is the potential for you to decide, not the decisions you make. For example, I could be walking through the woods and suddenly be presented with a choice of two paths. The path I choose is influenced by my self–images, past experiences, beliefs and philosophies. I might select one path over the other because of a million different self–identities, conscious and unconscious reasons. I may pick one path because it seems to be safe or the other because I like danger. So the choice I make and the self (me) are really intertwined. They are conditional and affected by the past—therefore not free will.

There have been physiological experiments in which a participant's brain is connected to a machine, and told to decide, through free will, to make a choice. The subject is instructed to choose a particular action, such as raise either their right hand or left hand. It's their free choice. Researchers discovered, however, that the brain signals the choice the person will make on an unconscious level well before the person consciously makes their decision.

To many, this experiment proves that free will does not exist. In truth though, the experiment is really measuring the subject's identity and the decision—not the spirit and free will. The decision, as well as the subject making the decision, are conditioned and influenced by past experiences. So naturally, this experiment would find that there is no independent self, making free conscious decisions, because the act of deciding is not separate from the self (image).

Free will is the unconditional potential to decide. It is present when your mind is quiet.

> *"To a mind that is still the whole universe surrenders."*
>
> *–Lao Tzu*

Any decision you make is conditioned by past knowledge and experience. Adam and Eve lived within a state of free will. They were present, free, and without need or want, therefore there were no outside or past persuasions on deciding. But when they became tempted to eat the fruit, they lost free will. The will became bound and conditional. By its very nature, free will is independent of the laws of linear time, physical change, and cause and effect. It is an inner quality of spirit which keeps us from existing like a robot, only capable of conforming to a past program. So the concept of free will is not a reality that can be proven scientifically, because it is not a material manifestation, but a spiritual gift. Genesis 1:5 describes the change or shift in awareness from an inner reality to an outer creation. However, it does not reveal forms changing. The power of time or change is embodied, reflected, and expressed in the 7th day.

Forms Changing (7th Day)

There is a sacred connection between time or change (Gen 1:5) and form (7th day), which allows for the generations of the heavens and earth and all life.

1 Thus the heavens and the earth were finished and all the host of them.
2 And on the seventh day God ended his work which he had made; and he rested on the seventh day from all his work which he had made.
3 And God blessed the seventh day, and sanctified it: because that in it he had rested from all his work which God created and made.

4 These are the generations of the heavens and the earth when they were created, in the day that the lord God made the earth and the heavens, Gen 2:5 And every plant of the field before it was in the earth, and every herb of the field before it grew: for the lord God had not caused it to rain upon the earth, and there was not a man to till the ground. Gen 2:6 but there went up a mist from the earth, and watered the whole face of the ground. 7 And the lord God formed man of the dust of the ground, and breathed into his nostrils the breath of life; and man became a living soul."

The 7th day is about the creator establishing an interaction and relationship between time and change with the heavens, earth, energy, nature, environment and body through living souls.

The 7th day reveals linear time, forms changing, and cause and effect. The potential of everything that God created in the 1st 6 days is realized through linear time on the 7th day. The heaven and the earth that we experience and all the things in it are quantifiable impressions of God's creation.

> *Mathew 24:35 "Heaven and earth will pass away, but my words will never pass away."*

Time in God's 6 days of creation is not subject to any laws, limits or dimensions. God can perceive time backwards or forwards, fast or slow, or any other way imaginable because it originates from within the unlimited light of spirit. The heavens and earth from the viewpoint of the 7th day exist within the dimension of a past, present and future. They will pass away.

On the 7th day, when God rests, he permits time to become entangled with us and his creation. This creates a sense of a linear process to change. Thus we get the generations of the heavens and the earth.

You can relate to the 7th day by recognizing that you hold the power of free will. Yet, you have a body, and live within an environment that is constantly changing.

When God's spiritual creation, Days 2 through 6, is perceived from the viewpoint of our physical brain, it becomes what we call *the universe*. Everything that seems to divide us from the universe, nature, and each other like distance, the past and future, thought, matter, and laws are not the essence of God's authentic creation. They are how our brain interpret, experience and connect to it.

From the creator's viewpoint, creation is finished. *Genesis 2:1 "Thus the heaven and earth was finished and all the host of them."* Yet, when creation is viewed through linear time, it appears to be unfinished. The process of creating, as we look into the natural world, seems to never end. New stars and life forms are being born every day.

Everything God creates is whole, eternal, and complete. This statement is aligned to the nature of God. The universe we experience through forces is an incomplete perspective of the whole of creation. This is why the creation of our universe seems to be continuously ongoing and uncompleted.

In the 7th day, spirit, with its attributes of light, consciousness, mind, perception, and free will, becomes individual souls in human form.

The universe is in an endless process of creatively imitating God's 5 outer creative acts on the Days 2 through 6. Our individual souls are in an endless process of seeking the 6 inner divine acts on the first day. It is through the 6 inner divine acts that our soul holds an impression of God's nature.

10

The Viewpoint Theory

The Search for Meaning

"If the whole universe has no meaning, we should never have found out that it has no meaning..."

—C.S. Lewis

The divine order offers a possible glimpse into how God created the universe and us, but it does not reveal why. Why are we here? Why did we end up with this reality or world? Science and religion tend to answer these questions very differently. Science explains our human existence and condition primarily through evolution and natural selection. Christianity understands it through the Adam and Eve story. Is it possible for these two opposing explanations to co-exist?

I don't believe that sorrow, disease, shame, fear, hatred, war, and hunger were part of God's intention for humanity. The question is then, which inevitably arises, if God did not intend for humanity to suffer, then why do we?

The Adam and Eve story illustrates that reality is determined by our relationship to God and our response to his creation.

> *"The physical world, including our bodies, is a response of the observer. We create our bodies as we create the experience of the world."*
>
> *–Deepak Chopra*

The Adam and Eve story describes spirit entering the world and becoming individual souls. The relationships between God, the soul, free will, consciousness, human body and environment set the stage for this drama.

Does the mind and consciousness determine the reality we experience? The Adam and Eve story seems to be based in the world behaving as a reflection of their consciousness. Adam and Eve initially lived in union with God. So the outer environment mirrored the quality of their inner relationship.

In the paradise, the Garden of Eden, the tree of life symbolized the perfect nature of God. The tree of the knowledge of good and evil signified the relative imitation and opposite of God's nature. If Adam and Eve ate from the tree of life, they would live aligned to spirit. They would be connected to God, and experience the sacred creation. But if they ate from the tree of the knowledge of good and evil, they would become detached from God. As a consequence, they would become vulnerable to the temporal environment of the natural world.

As the story goes, they were seduced by the snake (Satan), which tempted them with a promise—that if they ate from the fruit of the knowledge of good and evil, they would become like God. We all know this story has a sad ending for all of us. They chose to partake from its fruit. Now they were separated from God, spirit, truth, and essence. Therefore, they became exposed to physical laws and limitations.

Adam and Eve originally existed in what we might think of as a transcendent state of consciousness—fully present with God. It reflected a timeless reality. They had no reference to know they were naked, and so there was no shame or any possibility of disease or injury. In this transcendent state, their bodies or forms were not subject to the natural processes of the world. Their bodies were renewed instantaneously through their union with eternal spirit, and so there was no aging, physical or emotional suffering, and death.

After they ate from the tree of the knowledge of good and evil, they descended into our normal intellectual state of consciousness. Their "eyes were open" to a different reality. In this limited and relative world, they became part of the processes of time, cause and effect and change. Consequently, they were now susceptible to emotional and physical suffering, aging, and death.

I imagine many people believe that the Adam and Eve story cannot possibly have any kind of scientific basis, but that may not be true.

> *"The emergence of the arrow of time is related to the ability of the observers to preserve information about experienced events. Thus, the process of aging is related to our ability (or perhaps disability) to remember. A "brainless" observer, for example, would be able to not experience time or aging."*
> *—Robert Lanza M.D.*

Adam and Eve originally had minds that were completely present with God. So they lived in an eternal reality. When they experienced the knowledge of good and evil, it began the arrow of time for them, because now they processed reality through the memory of this knowledge. Aging then became part of their new reality.

Unfortunately, the Adam and Eve story does not end with them. It proclaims that this one act of Adam and Eve eating the fruit had dire consequences, for it corrupted or cursed human kind for every generation to follow. The knowledge of good and evil was symbolically or, perhaps literally, infused into humanity's DNA. It was passed down like a virus through the generations to every person, including you and me. It became part of our human memory.

It has been said that we live in a fallen and evil world. Can God's authentic creation become corrupted?

> Isaiah 43:13 *"From eternity to eternity I am God. No one can snatch anyone out of my hand. No one can undo what I have done."*

The Viewpoint Theory advocates that God's original creation remains eternally new, perfect and incorruptible, but our perceptions of it, within the generations, became tarnished through a diseased and depraved knowledge. Our world is not God's unlimited creation. It is the restricted impressions of it. Our limited ego or self-image is projected into our relationships with nature and each other and becomes the world.

> *"Quantum consciousness makes us realize that, being one with others and with nature, what we do to them, we do to ourselves..."*
>
> *–Ervin Laszlo*

So why would a benevolent and loving God allow the world to fall? In the 6 days of creation, everything conforms to God's viewpoint. In God's creation, there are not personal experiences and choices separate from God. On the 7th day, God allows being to live within a body that can interact with nature and time or change. This creates separate minds and souls. To fully be an individual soul, there must be free will, which means that no possible reality is denied, not even evil.

Our universe is the generations of God's creation. It is the memories and impressions of the 6 days described in Genesis. When Adam and Eve ate the fruit of the knowledge of good and evil, it altered their memory and consciousness of God and his creation. Their new memory of creation contained limitations based in a much longer history, perhaps billions of years. It was a memory of animals and humans struggling to survive, destruction, and death. This is the world that our science and history books describe. It is the reality of life and death, suffering, survival of the fittest, natural selection and evolution. It is a universe that seems unintentional, unfair and meaningless.

> *"The universe we observe has precisely the properties we should expect if there is, at bottom, no design, no purpose, no evil, no good, nothing but blind, pitiless indifference."*
>
> *–Richard Dawkins*

Most of us have heard the principle, credited to the ancient Greek philosopher, Heraclitus, that the only constant in the universe is change. We experience all forms in the universe, including our physical bodies, changing. The Adam and Eve narrative suggests that they initially experienced reality through forms which did not adhere to the natural world. Their bodies did not age and suffer. They perceived the universe from the inside- out. So all change moved through them. Their consciousness was the cause of all change in their world. When they ate from the tree of the knowledge of Good and Evil it reversed this dynamic. Now the outer world changed their inner consciousness. As a consequence, they became subject to and affected by, the laws of physics, time and biology. This made them vulnerable to degeneration, decay and entropy.

In the Abrahamic religions, Adam and Eve are believed to be the first humans on earth. There is a glaring disagreement between when scientists say the first humans appeared on earth versus when theologians claim Adam and Eve lived. Science suggests the first humans existed millions of years ago. Adam and Eve are believed to have lived thousands of years ago. This discrepancy seems to be an impossible problem. However, the focus on when Adam and Eve existed may be completely irrelevant. Time is a part of consciousness. The knowledge of good and evil changed the foundation of human consciousness. Therefore, it changes how time is perceived. If you can affect consciousness on a fundamental level, you can change your experience of the universe.

The universe will tell whatever story it has to for reality to align to our consciousness and viewpoint. The story that science tells of a big bang, evolving and expanding universe conforms to our fragmented and limited viewpoint. This is not the same story that was initially within the consciousness of Adam and Eve.

Evolutionist believe that what the Bible calls sin, like greed, envy, wrath, lust, and pride are simply evolutionary responses for survival. This is the gospel according to evolution and the natural world. But, the universe is simply reflecting within the limited realm of physical laws, causes and effects, and linear time, the nature of our selfish consciousness.

> *"Let us try to teach generosity and altruism, because we are born selfish."*
>
> *–Richard Dawkins*

The Adam and Eve story is really about the changing of reality itself. If you change your inner being or nature, you change the outer nature of the world. If you affect the observer, you affect the observed. So the story of evolution has a possible twist! It is the Adam and Eve narrative which caused the story that evolution tells. This narrative of habitual sin and suffering is a tragedy that we are all stuck in.

The Good News

> *Romans 6:23 "For the wages of sin is death; but the gift of God is eternal life through Jesus Christ our Lord."*

The divine order revealed that consciousness, or the impression of being, is an inner foundation of creation. It is more fundamental than time.

> *1 Peter 1:20 "He was foreknown before the foundation of the world but was made manifest in the last times for the sake of you."*

If you can affect consciousness on a fundamental level, you can change the past and future of the world. The Adam and Eve story affected consciousness on a fundamental level but so does the story of Jesus Christ.

> *John 16:33 "I have told you these things, so that in me you may have peace. In this world you will have trouble. But take heart! I have overcome the world."*

I believe Jesus established within human consciousness the principle of grace through the sacrifice on the cross.

> *Ephesians 8 "For it is by grace you have been saved, through faith—and this is not from yourselves, it is a gift from God."*

Through the power of grace you can transcend the limited experience of yourself–consequently receiving greater and more beautiful impressions of creation.

> *John 14:12: "I tell you the truth, anyone who believes in me will do the same works I have done, and even greater works, because I am going to be with the father."*

Consider this question: In the Adam and Eve story, did God change his view of them or did they change their perception of God? God does not change! After Adam and Eve ate the fruit of the knowledge of good and evil, their experience of God was changed from an unconditional loving being to a God of conditional love, judgment, and punishment. Perhaps this is why the nature of God in the Old Testament seems very harsh. The human relationship to God was based in the duality and limitations of the knowledge of good and evil. Jesus came to restore the true nature of God into the hearts and minds of humanity.

> *John 13:34 "A new commandment I give you, that you love one another, just as I have loved you; that you also love one another."*

God places no restrictions on free will, not even how we perceive him. It seems that many of the characters in the Old Testament experienced God as punishing, unforgiving and very intolerant. Yet, Jesus offered a vision of God which was infinitely loving, forgiving, and compassionate. How do you decide to see God?

> *1 John 4:8 "Whoever does not love does not know God, because God is love".*

Many people from all around the world, who have had near death experiences, often report seeing and feeling the light and love of Christ, spirits, colors, trees, animals, the earth and universe in a far more magnificent, loving and spiritual reality than we can even imagine. Some neuroscientists try to explain this away, as simply chemicals in the brain manufacturing images and experiences from past memories. The Viewpoint Theory suggests that the near death experiencer is, in truth, perceiving a greater potential of God's creation. Their impressions of creation are observed through a spirit light viewpoint, free from the past and our human sensory perception.

> *2 Corinthians 5:17 "Therefore, if anyone is in Christ, the new creation has come: the old is gone, the new is here."*

So maybe and simply, the meaning of our physical life is to be an individual human soul within the generations of the heavens and earth. The soul is like a conduit between God's creation and the natural world. We become a soul in a human body in order to learn, grow and love through personal experiences and relationships. Then, our physical body's form, experiences, nature, and relationships are imprinted within our soul. But, because our human bodies and brains contain a corrupted or unholy memory, the soul cannot ascend this earth. Fortunately, the memory of Christ's resurrection is also within us and overcomes the weight of sin.

> *Romans 8:11 "The same power that raised Christ from the dead is living in you."*

So through the Christ, our soul can carry the human body's memories and impressions back with us, after our death, to our true home, closer to God and his creation.

> *John 14:2 "My father's house has many rooms; if that were not so, would I have told you that I am going there to prepare a place for you?"*

11

The Viewpoint Theory

If God Created the Universe

"Faith doesn't make sense that's why it makes miracles."

–John Di Lemme

To believe or not to believe in God—that is the big question. Does the universe possess the answer? There are many assumptions about what reality would be like if God created the universe. Here are a few:

– If God created the universe then it must be intelligently designed.
– If God created the universe then there must be evidence and proof.
– If God created the universe science would know.

Modern science has done a masterful job in offering natural causes and explanations for our reality through the fields of physics, quantum physics, biology, and so on. Still, some argue that the finely tuned nature of the universe points to an intelligent designer.

God is more than his creation, and his creation is more than the universe. Both natural cause and intelligent design are human perceptions of the universe. The supernatural creator allowed the natural world to manifest effortlessly, or by resting. This is why our universe appears to be caused by and function through natural processes. Without God letting the natural world form, would we have a soul and free will?

> *Genesis 1:31 "and God saw everything he had made and behold, it was very Good."*

The word *Good* in Genesis is not to be taken lightly or casually. We use that word indiscriminately for anything that is pleasing to us like, "You did a good job," or "That was some good food." Its meaning in Genesis though, is far more significant than that. The word *Good* embodies and reflects the whole nature of God. So it embraces absolute love, omniscience, omnipresence, omnipotence, perfection, and so on. Jesus clarified the amazing significance of the word *Good* in *Luke 18:19 "Why do you call me good?" Jesus answered, "No one is Good except God alone."*

God's creation is synonymous with Good. There are certain words naturally inherent to a creator, such as *spiritual, supernatural, perfect, miraculous, instantaneous, eternal, unlimited, free,* and *infinite*. These words are distinctly different from the words *intrinsic* and *relevant* to traditional science. The words and concepts such as *physical laws, forces, chance, randomness, measurable, predictable, finite, limits, adaptation, natural selection, evolution, and probabilities* are not relevant to God's creation.

Create versus Design

Intelligent design is a belief that the universe could not have just happened by chance or natural causes. Therefore, there must have been an intelligence that designed it. The word *Create* applies to God's 6 days of creation. *Design*, as we think of it in terms of science, does not. It is relative to the 7th day or our universe.

Is the identity and definition of a designer intrinsic to God's nature? God is a creator! The creator, by his very nature, needs nothing outside his own existence to create. There are no rules, mathematics, science, limits, measurements, regulations, and laws to abide by in order for God to create anything. This principle is emphatically demonstrated in *Genesis 1:1 "In the beginning, God created the heaven and the earth."*

The notion of a designer implies that there are rules and mathematical principles that must be followed. *Design* is not a characteristic of God's creation, unless you can define that word in a spiritual way beyond all science. We associate *design* with the material world, where everything is definable and has measurable dimensions and limits.

God, as a *designer,* is a concept built upon this tangible and finite world. If you set out to build a house, the first thing you might do is design it to your liking. Then you would calculate and figure out all the measurements and dimensions of your design. After that you would go buy the materials and tools needed to construct your house. Finally, you would begin its construction, and follow the measurements of your design until you were done.

The method of how we would construct a house is equated to how God would create the universe.

> *"The whole difference between construction and creation is exactly this: that a thing constructed can only be loved after it is constructed: but a thing created is loved before it exists."*
> —Charles Dickens

Contemplate this insight for a minute: if an eminence of God's nature is omniscience (all knowing), there would be nothing to think about or figure out—no measurements to take, no math problems to solve, no theories to form, no tools needed, no laws to abide by and no experiments to test. Therefore, no material design is needed for God to create the universe.

When we read the Genesis creation, we are prone to mistake God's intentions as designs. For example, on the 4th day:

> *Genesis 1:14 "And God said, let there be lights in the firmament of heaven to divide the day from the night; and let them be for signs, and for seasons, and for days, and years."*

Some may view this verse as God designing the universe. What we see as designs, measurements, and spans of time are the impressions of God's creation. God's intentions and creations are complete, absolute, whole and immeasurable. From our viewpoint, however, they become relative, measurable, and quantifiable through time. So *design* is our limited interpretation of God's intentions and creations.

God's intentions for days and years, we assume, represent 24 hours and 365 days respectively. Astrophysicists claim that a day and a year on a planet across the cosmos could be the equivalent to a thousand years on earth. The relativity of time is an effect of the 4th day in Genesis. God's incomprehensible creation of days and years could be experienced as countless different spans of time from a human viewpoint.

> *2 Peter 3:8-9 "…With the Lord, a day is like a thousand years, and a thousand years like a day."*

Deliberate Creator versus Natural Science

In the great debate between atheists and intelligent designers, the natural sciences are used as the judge and the jury. Science can determine the fate of whether a creator exists or not. The natural world is viewed as an impartial and unquestioned revealer of all truth. So the belief is that all reality passes through the mediums of the natural sciences.

From an unlimited viewpoint, the natural sciences are biased mediums through which only a partial reality is observed. Perhaps, this is why physicists encounter this principle of nothingness. The nothingness that they theorize somehow exists, is simply the limits of what the universe can offer as a reality.

In academic world, the universe and science are treated as absolute truths. God is seen as something to believe in or perceive. But, from a limitless viewpoint the opposite would be true. The creator and his creation are the absolute truth, and the universe and science are the beliefs or the perceptions.

> *"Traditional science assumes, for the most part, that an objective independent reality exists; the universe, stars, galaxies, sun, moon, and earth would still be there if no one was looking."*
>
> *–Deepak Chopra*

Some physicists are predicting that soon there will be the discovery of a theory of everything. They believe science is close to connecting all the aspects of the known universe. Will this theory of everything, really be everything?

From the viewpoint of a limitless creator, what physicists see as *everything* is really an impression or imitation of *everything*. The real *everything* is God's creation, which is unperceivable, infinite, unexplainable, unknown and immeasurable through any device or means.

Our universe mimics the whole or completeness of God's creation. This creates the illusion that the universe is self–contained, self–explainable and everything. If other universes exist, they would also be mimicking the authentic creation. There can be nothing greater than God's creation.

Evolution

The most heated argument between creationists and atheists concerns evolution. Evolution basically states that every living organism evolves from simple to complex, through gradual steps in time. This theory is at odds with the creationist view. They believe God created all life in its completed form thousands of years ago.

The Viewpoint Theory proposes that God did create everything in its finished and perfect form. Its flawless and completed form, however, is spiritual. It acknowledges the perfect nature of God. The universe we experience is a replication or reproduction of God's creation. Flaws are inherent in anything replicated. But the extent of flaw in the universe lies within the beliefs and mind of the observer.

"In a universe of blind physical forces and genetic replication, some people are going to get hurt, other people are going to get lucky, and you won't find any rhyme or reason in it, nor any justice."

–Richard Dawkins

Evolution, along with all natural occurrences, serves the process of creatively reproducing or replicating God's creation. It is a partial and time–bound natural response to the inability for our brains to fully comprehend the finished creation.

Our bodies are intertwined with physical and chemical processes. We have a physical sense of self that is interconnected with the generations of the heavens, earth, and life forms. All the complexities of life are guided by evolution and the forces of nature. So our human survival instincts, emotions, habits, experiences, fears, motives, desires and so forth, make up our physical consciousness. But they do not comprise our spiritual consciousness. Spiritual consciousness is eternally present. You awaken to it and receive it from God. It does not evolve from the past and physical world.

John 3:6 "Flesh gives birth to flesh, but the spirit gives birth to spirit."

Faith versus Evidence

In this new age of technology, there is an unquestioned assumption that facts and evidence are the masters of all reality.

> *"Faith is the great cop-out, the great excuse to evade the need to think and evaluate evidence."*
>
> *–Richard Dawkins*

Math concepts, such as probabilities, are viewed as neutral and immune to bias concerning any subject including God. They are perceived to have the unlimited capability to distinguish what is real from what is not.

What is the probability that God exists? This question may appear to be valid and neutral. But this very question fabricates a biased answer in favor of the non–existence of God. If God does exist, it means that there are levels of reality beyond the paradigm of probability.

Our universe accommodates the viewpoint of probabilities.

I once read a study which concluded that people who applied sun screen were much more likely to get skin cancer than people who did not. But what they failed to include in their research, for whatever reason, was that people who use sun screen are the ones who go outside and are exposed to the sun. So the study was biased and misleading because it did not include all the factors that may lead to skin cancer.

Science is not capable of factoring God's nature into reality. So trying to calculate the probability of God's existence, without the capacity to include God's nature, will lead to deceptive results. From a limitless viewpoint, all the methods of math and science are completely inept in capturing a creator.

Evidence

John 18:36 "My Kingdom is not of this world..."

Where is God's creation? It is here and now! But what we see and science finds is the evidence of it. Our world is evidence. The material universe and evidence are the equivalent. The physical world is not God's creation; it is the evidence of it.

Romans 1:20 "For the invisible things of him since the creation of the world are clearly seen, being perceived through the things that are made..."

Evidence of God's creation cannot be isolated to any time or place. Proof won't be exposed by digging it up like dinosaur bones or looking through a new powerful telescope. Evidence of God's creation is in everything that is knowable.

Evidence is a quality of the imitation, or replication, of creation. It is not an eminence of God's authentic creation. The limited nature of evidence exists within our human foundation of reality: space, linear time (past, present, and future), cause and effect, and physical change. Our universe is not capable of apprehending through evidence and facts that which created it. Evidence only reveals a fragment of reality.

I once heard a skeptic say that if God showed up in the world, science would know it. But this statement discounts God's unlimited nature. *The Viewpoint Theory* suggests that how God interacts with the material universe is a mystery to the universe itself. The world we perceive is incapable of recognizing the miraculous and supernatural.

> *John 1:10 "He was in the world, and though the world was made through him, the world did not recognize him."*

If miracles happen! They are a paradox for science, because by their very nature they cannot be replicated, and replication is the nature of our universe. Our universe, how we currently perceive it, based in linear time cannot record the timelessness of a miracle and then replay it. Therefore, miracles cannot be proven, so scientifically, they don't and can't happen.

Faith

No matter if you believe God exists and created the universe or not, ultimately, everything still comes down to faith.

> *"... over the entrance to the gates of the temple of science are written the words: 'Ye must have faith.'"*
> –Max Planck

Faith itself is the spiritual evidence of God.

> *Hebrews 11:1 "Now Faith is the substance of things hoped for, the evidence of things not seen."*

Acknowledgements

There are many people that have helped make this book possible. Without these special few, it would have been impossible.

Special acknowledgement goes to Bill Lefko for his years of guidance and friendship and for his wisdom into the nature of reality.

Special thanks to Clifford Burns for his creative thinking and ability to see outside the box and gain unique insights into the universe.

Special thanks to Bruce Donaldson for his endless support and friendship and for his unconditional giving nature.

Special thanks to Stacie Williams for her belief, persistence, and help in making this book a reality.

Special thanks to David Bedenbaugh for his challenge on the tennis court, support and Christian perspective on creation.

Special thanks to Jim Willis for the selfless gift of his time, energy, guidance and wisdom.

A Very, very special thanks to my fiancée, Lorrie Moneymaker, for her love, patience, and encouragement

And to our grandchildren, Christopher Crosby, Aaligha White, and Triston Cole.

www.ingramcontent.com/pod-product-compliance
Lightning Source LLC
Chambersburg PA
CBHW060841220526
45466CB00003B/1199